EPOXY BASICS

Working with Epoxy Cleanly & Efficiently

2nd Edition

by
Russell Brown

EPOXY BASICS

Working with epoxy cleanly and efficiently.

Photos and text by Russell Brown of Port Townsend Watercraft
Edited & cover design by Ashlyn Brown
Photo contributions (*) by James Island

This book was originally written as an addendum to our construction manuals for small boat kits produced by Port Townsend Watercraft. People noticed and asked for copies of just the addendum, thus we created an e-book and hard copy versions. This book is not intended to be a complete "how-to," but it does give a pointed overview of working with epoxy, describing methods used by the author for, Gluing, Filleting, Glassing, Coating, and other epoxy techniques.

This revised, Second Edition (2026) has many small changes. Yes, our opinions change and we do seem to learn new things over time.
In this book, there are some personal preference references to various products and companies. These are offered as a convenience only and are not intended to exclude other companies that might offer similar products.

A big *THANK YOU* goes to all of our customers over the years, our friends, and associates.

Russell and Ashlyn Brown started their kit boat business in 2009. Their company, Port Townsend Watercraft, aims to encourage rowing/sailing tenders (especially nesting dinghies) as an alternative to inflatable tenders. Learn more about their kits, small boat solutions, and other maritime books, at ptwatercraft.com and on PT Watercraft's blog.

TABLE OF CONTENTS

EPOXY

Epoxy technology is relatively new. As such, it's not as much of a taught art as other building methods. An art it is though, and there are many techniques for using these materials efficiently.

This book can help one to work clean and fast, minimize sanding, conserve materials, and most importantly, how to build something that is as strong and long-lasting as it can be for its weight. It covers, in brief, the four basic methods used in wood/epoxy boatbuilding; gluing, glassing, filleting, and coating.

To achieve the full potential of these materials, one needs a warm and dry place to work.
The viscosity of epoxy is greatly affected by temperature, so mixing, coating, and glassing are faster and easier at comfortable room temperature. Cure times are faster as well and this helps overall progress.

If working in winter, most heat sources (not Propane) will help keep your workplace dry.
If working in a damp environment, a shop that can be closed to keep out night air, fog, etc., is recommended. Electric dehumidifiers (the kind used in basements) work great for keeping a shop dry.

MIXING A BATCH OF EPOXY

This seems pretty simple, and it can be, but it's a deep subject. Epoxy is a two part, fairly viscous material that needs to be **metered accurately** and **mixed thoroughly.**

The epoxy that we recommend is WEST SYSTEM®. We use this epoxy for its excellent physical properties (strength) and reliability. Our experience has been that WEST SYSTEM® has always shown responsible product testing, honesty in marketing, and superior products.
WEST SYSTEM® also has a tech support team, based on over 50 years of experience with epoxy, ready to tackle just about any question you may have. Just give them a call.
WEST SYSTEM®'s literature, no matter what epoxy you are using, is recommended reading. Always refer to the product information for the epoxy you choose.

If you use WEST SYSTEM® epoxy, there are a few hardener choices to use with the **105 RESIN®**
Depending on the size of the project and working temperatures, **205 FAST HARDENER®** and **206 SLOW HARDENER®** are standard choices.
207 SPECIAL CLEAR HARDENER® is great for areas which are to be clear finished, but can also be used for the whole project. **The advantages** of the 207 SPECIAL CLEAR HARDENER® are: It doesn't produce *amine blush* (pg 40), it is slightly thinner, which is helpful for glassing and coating, it is clear for bright finishes, and it has similar physical properties to 205 FAST HARDENER® or 206 SLOW HARDENER®.
The cure time with the 207 hardener is somewhere between the fast & slow and it is a bit more expensive.
207 SPECIAL CLEAR HARDENER® mixes with 105 RESIN® at a 3:1 ratio (by volume).
205 FAST HARDENER® or 206 SLOW HARDENER® mix with 105 RESIN® at a 5:1 ratio (by volume).

Epoxy must be metered accurately to the correct ratio.
Epoxy metered inaccurately may get hard, but it will not be nearly as strong as it would be if it was mixed to the prescribed ratio.
Never add extra hardener, and always discard a batch that you are not sure about.

The options for metering WEST SYSTEM® epoxy can be seen in their catalogue, or see our ProBoat article about mixing & metering epoxy; https://ptwatercraft.com/blog/mixing-metering-epoxy/
Mini pumps are a good option as they are inexpensive and reasonably accurate. These pumps fit the jugs that the epoxy comes in. If you decide to use mini pumps and the temperature in your shop is not always warm, you will need to keep the epoxy jugs warm. Once primed, it's one full stroke from each pump to get the correct ratio. When more than one pump of epoxy is needed, always start with a pump of resin, then a pump of hardener and repeat, instead of trying to count pumps.
Note: *Some epoxies use mini-pumps that require two or three pump strokes of resin to one pump of hardener, which can be confusing.* ***Read the manufacturer's directions.***

Keeping the epoxy warm will do two things; it will greatly increase the speed of pumping for mixing large batches, and it will make the two parts easier to mix thoroughly together.
A drop light with a 40 watt bulb clamped close to the resin jugs or pump is one way to keep the epoxy warm.

CUPS AND MIXING STICKS

You will likely be mixing many batches of epoxy. It helps a lot if you get the right cups and sticks, and learn how to do this part well.
We think of epoxy as liquid gold. Properly mixing it is like adding value.

To mix all the epoxy in your cup, you will need cups with straight sides and a flat bottom, and stir sticks with straight edges and square ends.
With this combination you will be able to scrape the unmixed epoxy off the sides and bottom of your mixing cup, so that your whole batch gets mixed together thoroughly.

We use lots of 12 oz paper coffee cups. If you buy coffee in paper cups, wash and dry the cups. Get your friends to do the same, and then you may only have to buy the quart size paper cups. Quart sized cups will be used for larger glassing, coating, and filleting operations.

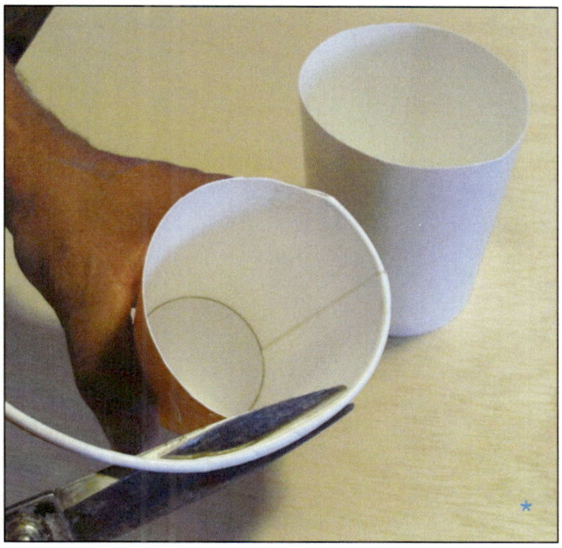

Try snipping off the tops of the cups with scissors.
This makes a sharp edge you can scrape your mixing stick and other tools on, keeping epoxy inside your cup, and not running down the outside. Try it, you'll like it.

Stir sticks of different sizes can be cut on the table saw. They must be wide enough to get the batch of epoxy moving, but light (thin) enough to not tip the cup over when it is empty.
The edges of the sticks should be straight and the ends square.

We always clean our stir sticks and other epoxy tools carefully after use by dry wiping, or if they are really sticky, a bit of denatured alcohol on a rag or paper towel.
Sometimes the stir sticks will need to have their edges straightened on a sanding block.

MIXING

Hold the bottom of the cup in the base of your palm. Stir the two parts so that you get the whole batch moving around in the cup. Stir some more.

Now scrape all the unmixed epoxy from the walls of the cup with the stir stick held vertically. The edge of the stick should contact the walls of the cup all the way around.

Next, holding the stir stick horizontally, scrape the unmixed epoxy off the stir stick, and into the cup. Stir again, repeat the above process, and stir again.

Note:
Always thoroughly mix the two parts together *before* adding fillers.

EXOTHERM

This is what happens if that batch of epoxy stays in the cup for too long before being used. It gets hot, maybe even smoking hot, and gets hard. This happens only when mixed epoxy is in sufficient mass or volume to heat itself up.
To avoid exotherm, get the epoxy out of the cup and onto the job as quickly as possible.
When a batch of epoxy starts to get warm, it can be a big mistake to continue using it.
For large epoxy jobs, preparation is the key to success.
If a cup of epoxy starts to exotherm, place it outside, as the fumes could be harmful.

GLUING

Gluing is a multiple step process. In brief: **Parts to be glued are usually given a coat of un-thickened epoxy, or "primed" on both surfaces. Fillers are then added** (to the same batch, or to a new batch of epoxy), and this **thickened epoxy is applied to one surface** of the parts to be glued. The parts are assembled, and the squeezed out glue is cleaned up.

Priming is a very important step. Saturating the surfaces to be glued with un-thickened epoxy before applying thickened epoxy improves bond strength and helps prevent a "dry" joint.
Priming even helps adhesion on pre-coated or glassed surfaces, as the thin epoxy will better wet through residual dust and key into the sanded surfaces.

In areas where parts with end grain wood are being glued (such as a butt joint or scarf joint), apply un-thickened epoxy multiple times to let the wood absorb as much as it can before applying the thickened glue.

The prime coat is most often applied with **a foam roller** (see page 47) cut to an appropriate length.
Using a roller is much faster and cleaner than using a brush for applying epoxy. It is also usually more economical considering that one can cut the roller to the length needed and apply a thin, even coat without drips running down the edges. Cutting the section of roller as short as possible for the job saves epoxy.

FILLERS FOR GLUING

Fillers are mixed into the epoxy to make thickened glue. (numbers noted are WEST SYSTEM®)

406 Colloidal Silica® makes epoxy very friendly to work with when wet and very dense and hard when cured, but add some fibers (**403 Microfiber Filler®** or **404 High-Density Filler®**) to the batch of glue as well.

403 Microfiber Filler® is course and makes the glue a bit harder to work with, but is beneficial as the size of the fiber helps keep a film of glue between the pieces being glued.

404 High-Density Filler® is less coarse, easier to use and performs a similar function.

None of the above fillers actually make the epoxy stronger (with the possible exception of # 404), but thickening the epoxy is necessary for bonding most parts. **For strength, avoid over-thickening the epoxy.**

If using clamps for glue pressure, avoid over tightening them. If all the glue is squeezed out by over clamping, and the wood is still absorbing epoxy, a "dry" joint can result.

The thickness needed for glue depends on the job.

Some glue-ups cover a wide area and have minimal clamping pressure.

For jobs like this, one would mix glue thin enough that excess glue would still be able to squeeze out.

Other glue-ups need glue about as thick as catsup or mayonnaise.

Spreading glue over a large area should be done with a notched tooth trowel.

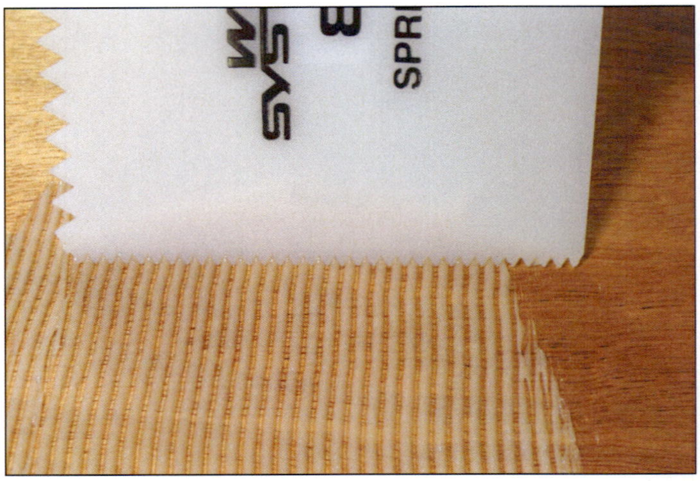

Applying glue for long, narrow glue joints, such as glue cleats or stringers can be done quickly with a *'squeezie' bag* (see filleting on pg 15).

The bead of epoxy applied with the *squeezie bag* can be flattened or spread out with a gloved fingertip.

Clean up is a critical part of gluing, unless you like sanding.
Colloidal silica filler (see page 8) makes the epoxy easy to work with when wet, but sanding it later is like sanding rocks, so sharpen your chisel sticks (described in the filleting section) before you get started gluing.
Time spent cleaning up squeezed out glue beats a lot of time spent sanding.

One of the best tricks ever for clean-up is the plastic straw. Snipped off at 30 degrees or so, the straw can clean up squeeze-out like nothing else!
Push a short distance, dab on a scrap of cardboard to remove glue, push again, etc. When mostly clean, wrap a paper towel around the straw and pull to clean, then make final passes that way. We use (***and re-use***) both large and regular size straws.

For gluing really critical joints, such as scarf joints for a spar, extra precautions should be taken to prevent a "dry" glue joint. A dry joint can occur when epoxy penetrates into the grain of the wood after the joint has been clamped together.

To encourage maximum penetration of epoxy before gluing, use a heat source to warm the wood **before priming** with un-thickened epoxy. **Cooling wood will draw the thin epoxy in**, especially when end grain is present, as it is in a scarf joint.
Warm the joints (*not too much*) with electric heat, heat lamps, or heat gun.

Add more un-thickened epoxy as it soaks in before applying the thickened epoxy.

Thickened epoxy is applied to one side of the joint only.

The choice of fillers for thickening epoxy for gluing is important.

Colloidal silica inhibits penetration, which is good because once the joint is clamped together penetration is to be discouraged.

Micro fibers (used in lesser quantities) helps to keep a film of glue between the clamped joints.

We have clamped this scarf joint between two flat boards (with a plastic or waxed paper barrier between) so that the joint can be made flat, but without applying any more clamping pressure than needed.

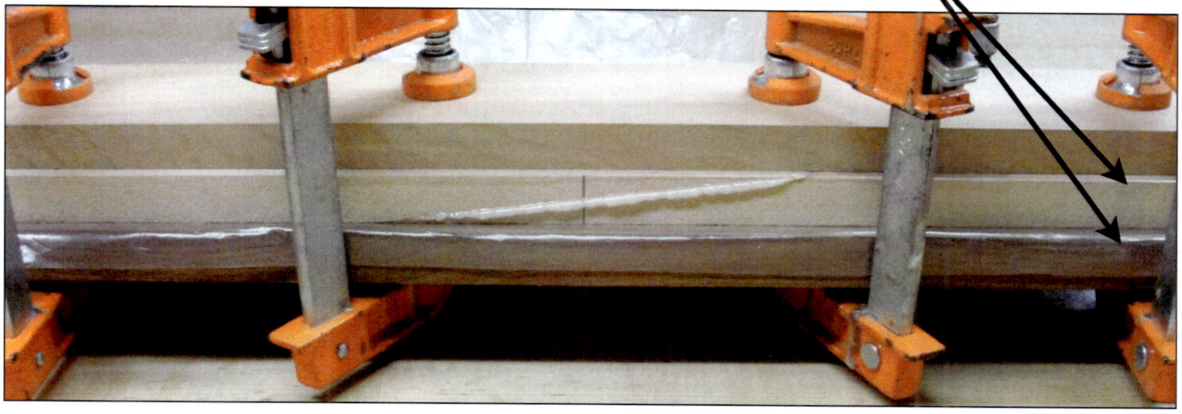

ALTERNATIVE CLAMPING METHODS FOR GLUING

We use drywall screws with plywood pads to achieve clamping pressure in many areas.
In the case below, holes were drilled in the hull side for the screws to pass through.
A dry run was done, locating the gunwales and driving the screws into place, so that the gunwales would locate correctly later when gluing.

In the case shown, we are using 1", # 6 screws.
The holes drilled in the hull sides are 5/32".
The holes drilled in the plywood pads are a bit bigger (3/16") to allow easy removal of the screws from the pads. Why? Because before gluing, the screws must be separated from the pads, sprayed with WD-40, and blotted on clean rags to remove excess oil.
All this to prevent the epoxy from bonding to the screws. Even so, the screws should be removed the following morning to avoid breaking the screws.

The better spray to use for screw removal is a dry-lube called *Mclube SailKote®*, which is a common sailing lubricant.
Lay the screws on a rag, spray them, roll them around and spray again.

For parts glued flat on the work bench, nails have several advantages over screws.
Besides providing alignment (more below), they leave smaller holes to be filled with epoxy (See syringing and color matching on page 42). If shiny smooth nails are used, there is no problem with them getting stuck in a glued up part.
A simple butt joint in thin plywood is shown being glued using small nails to hold it flat the the bench. *Plastic packing tape* was applied to the bench to keep the part from sticking, both edges were primed, and thickened epoxy applied to one edge with a squeeze bag before nailing the parts to the bench.

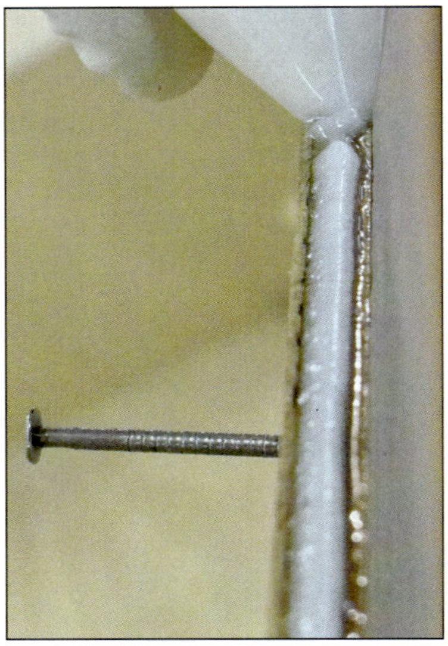

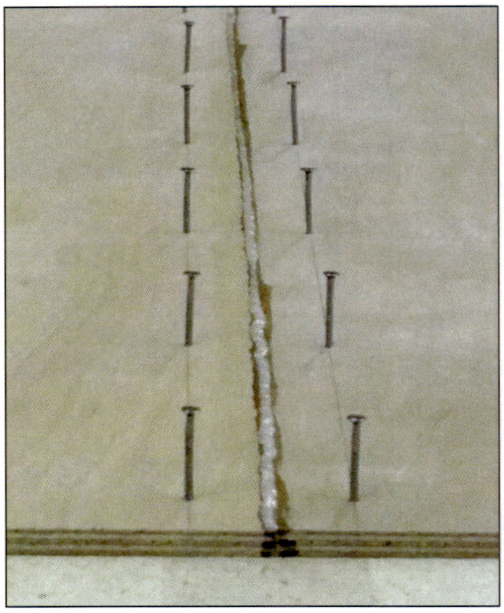

On the parts below, the nails were driven through small (1/2" x 1/2") plywood pads to achieve more gluing pressure. **The glue cleats shown below use the nails for glue pressure and for alignment.** The cleats have been pre-drilled with holes small enough that the nails need to be hammered in. **The cleats were positioned flush with the edges** and the nails driven through into the panel about 1/8" as shown. This allows the cleat to be immediately and accurately re-positioned when gluing.

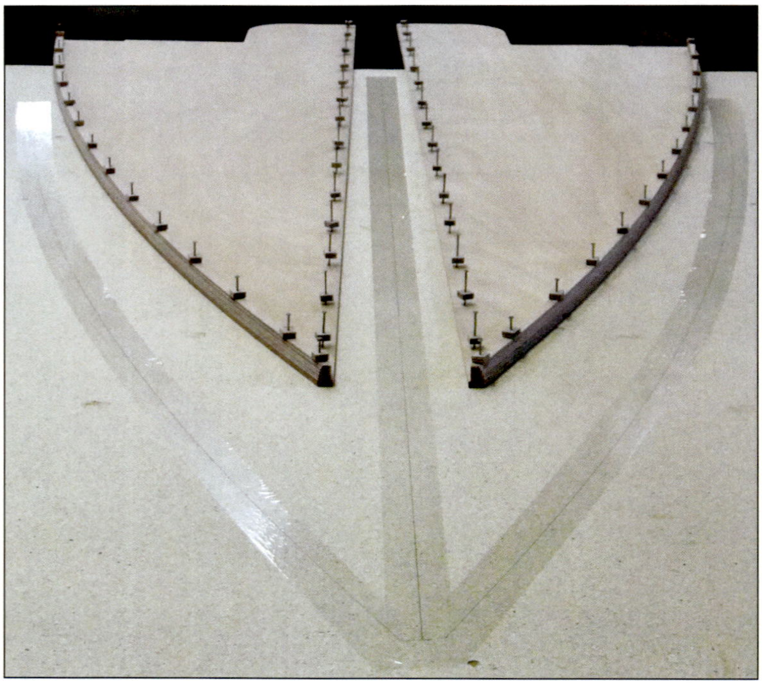

Below; The parts are primed, thickened epoxy is applied, and the cleat is positioned with the nail tips easily finding their holes before hammering the nails through the panel and into the bench.

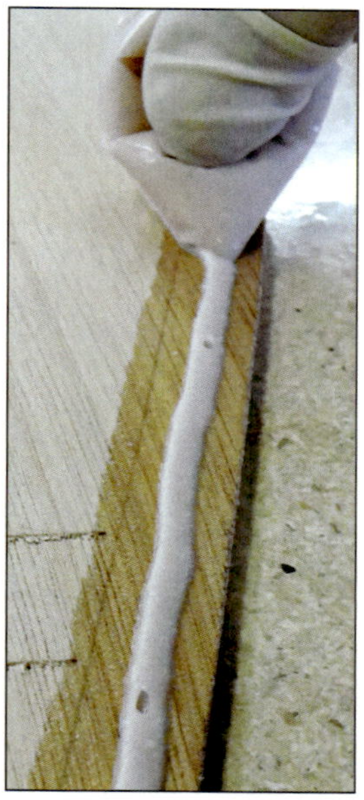

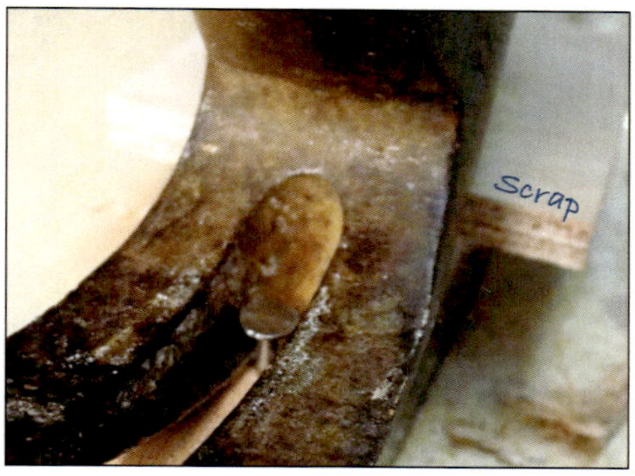

The nails are easily removed after the epoxy has cured by breaking away the small plywood pads with the hammer and pulling the nails with a scrap under the hammer.

We often use small nails (# 17 wire nails) for alignment of parts to be glued with clamps. Without alignment nails, the parts slide all over the place when gluing and accuracy is lost. The trick is to leave a bit (1/8" or so) of the nail protruding from one piece into the other so that the parts easily locate when gluing.

FILLETING

Filleting is like welding. A good weld is strong, looks good, doesn't require lots of grinding, and a good welder can do them fast.
Filleting is similar to welding in a structural sense and filleting is every bit as technical. Both materials are liquid when being applied, but the similarities stop there.

SURFACE PREPARATION FOR FILLETING (See surface prep notes on pg 40)

The strength of a fillet depends on a good bond. Most of the filleting that we do is on pre-glassed or pre-coated surfaces that have been sanded before assembly, but we still sand before filleting.

Sanding removes contaminates and the residue from gluing and it also removes the little glue bumps that the fillet stick could ride over and the chisel stick could catch on.
We use sandpaper that is glued back to back with an edge folded over to sand for fillets.
Vacuum up the dust with a small, clean, brush attachment and rub areas to be filleted hard with a clean rag to remove residual dust (again, see page 40).

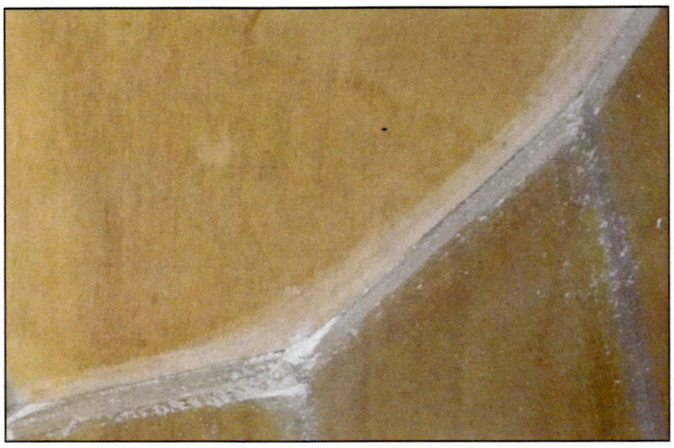

FILLERS FOR FILLETS

Fillers are mixed into the epoxy at the time of use to thicken the epoxy. Different fillers are used to produce different properties.

For structural fillets, we like to combine different fillers to create the properties we want.
406 Colloidal Silica® is necessary for workability of the thickened epoxy when wet.
407 Low-Density Filler® adds volume (extending the batch of epoxy) and makes the fillets easier to sand.
404 High-Density Filler® (used in lesser quantities) adds some strength without making the wet epoxy harder to work with.

These fillers can be pre-mixed (dry) in a gallon plastic bucket with a lid.
Measure the fillers with a cup: 2 cups colloidal silica, 2 cups low-density filler, and 1 cup high-density filler.
Put something heavy in the bucket, like large nuts or washers, put the lid on the bucket and shake.
The fibers may settle, so shake it up before use (with the lid on).

Avoid breathing fillers when mixing them into the epoxy by wearing a good respirator or by using an exhaust fan, both described on page 49.

FILLETING TOOLS

A set of fillet sticks can be made from thin plywood or can be purchased from PT Watercraft. These should be sharpened as shown with a sanding block and labeled. The numbers represent diameter (from 1/2" to 2 1/4") and correspond with fillet sizes called for in our manuals.

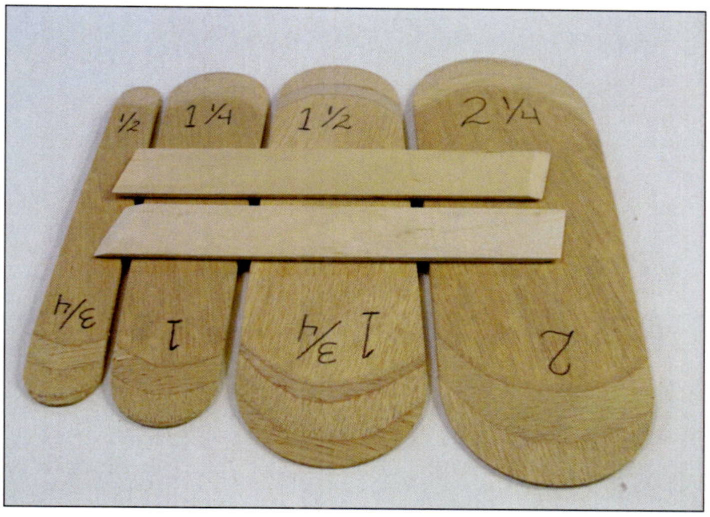

Chisel sticks, named such because they are sharpened like a chisel (and get shorter every time they are sharpened) are shown in both photos. **These work like magic for cleaning up epoxy and save countless hours of sanding, but they must be sharp to work well.**

Chisel sticks can be made from tongue depressor sticks (should be thin, light & flexible). Our favorite ones are made from 1/32" G10 fiberglass plate.
Both ends are sharpened, but done so on opposite sides for cleaning both ways.
A sanding block with really fine paper works well for sharpening these important tools. A burr will develop when sharpening, fold the burr over with your finger and make one last pass on the sanding block.

"Squeezie" bags are quart size ziplock bags and are used as described for applying thickened epoxy.
Freezer bags are the easiest to use and puncture resistant.
Fold the top of the bag over for loading.

When mixing fillers into epoxy for filleting, stir slowly to avoid losing valuable fillers. Use a larger cup for big filleting jobs and try to 'fold' the fillers into the epoxy.

Your thickened batch of epoxy should be thick enough that a large blob of it will hang on your mixing stick without drooping.

Pack the epoxy into the corner of the bag with a clean stir stick held flat.

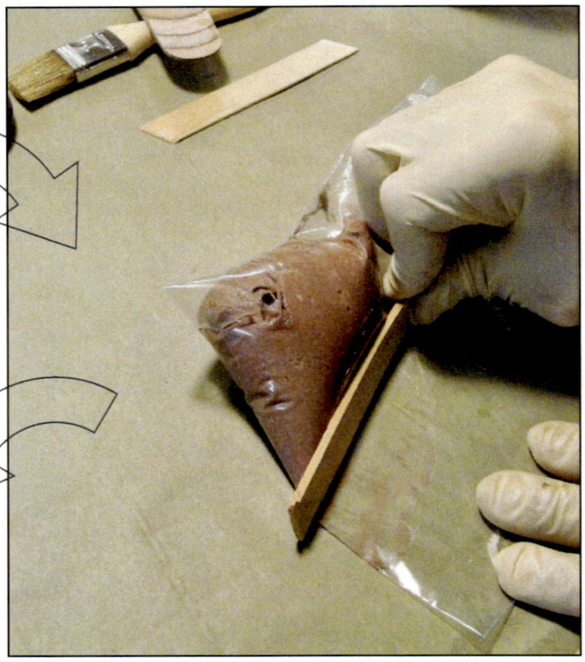

Gather up and pinch the bag at the epoxy, ...

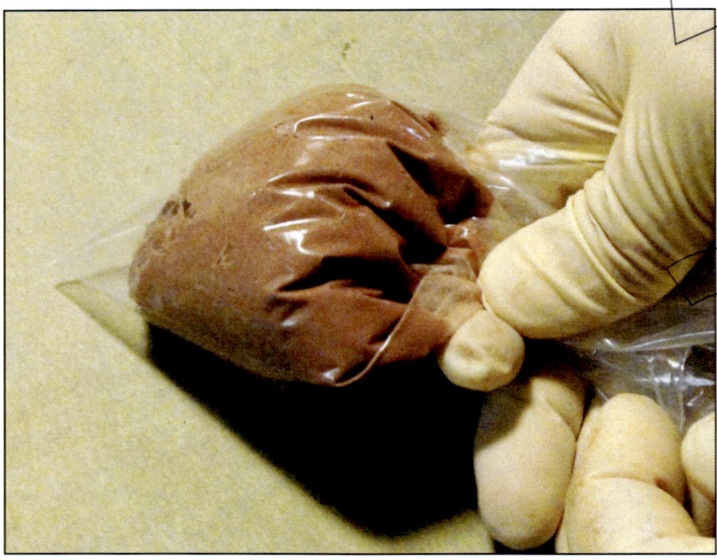

...twist it firmly, ...

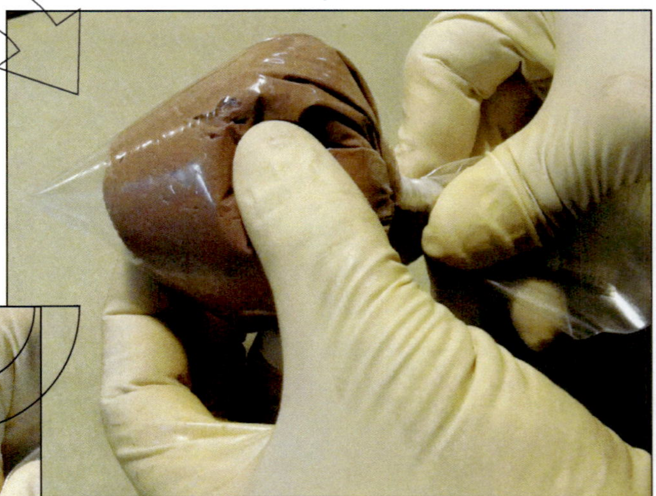

...and snip off the tip of the bag with scissors.

Ready to start filleting? Do you have everything ready? "Squeezie" bags, a 1" disposable brush, fillet sticks, chisel sticks, cardboard pallets for all your sticky tools, paper towels, etc.

Priming for fillets does two things:
It makes your bead of thickened epoxy slide into the corner with less chance of trapped air behind it, and it also makes your fillet bond much better.
It does this by penetrating bare wood, and in the case of secondary bonding, brushing the thin epoxy will lift up any residual dust on the surface, and help the liquid epoxy bond to the sanded epoxy.
Use just a little epoxy for priming and wipe up the excess with paper towel.
Proceed directly with filleting.

APPLYING FILLETS

Start with a short section. **Hold the tip of the "Squeezie" bag firmly in the corner to be filleted,** so that the bead of epoxy goes directly where it is needed without getting air trapped behind it.
Pull the bag, don't push it, and stop often to twist the bag to keep it firm, like a balloon.

Use the fillet stick to check that the bead of epoxy is appropriate for the fillet size.

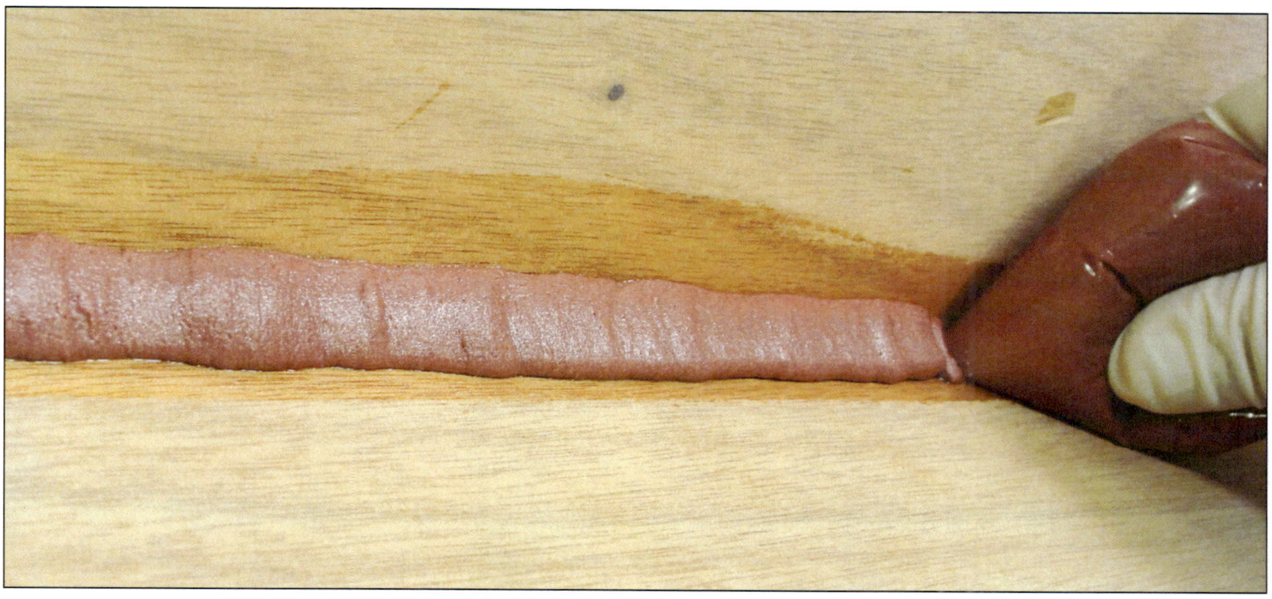

The volume of the final fillet depends on the angle at which you hold the fillet stick.

The way to do a really clean fillet is to apply enough epoxy that when the stick is at, say 50 or 60 degrees to the surface, it forms a fillet with only a few voids. Epoxy can be added to the low spots if needed, and on the next pass hold the stick at 80 or 90 degrees.

Clean up the excess with sharp chisel sticks. Chisel sticks drastically reduce sanding time, but they must be sharp.

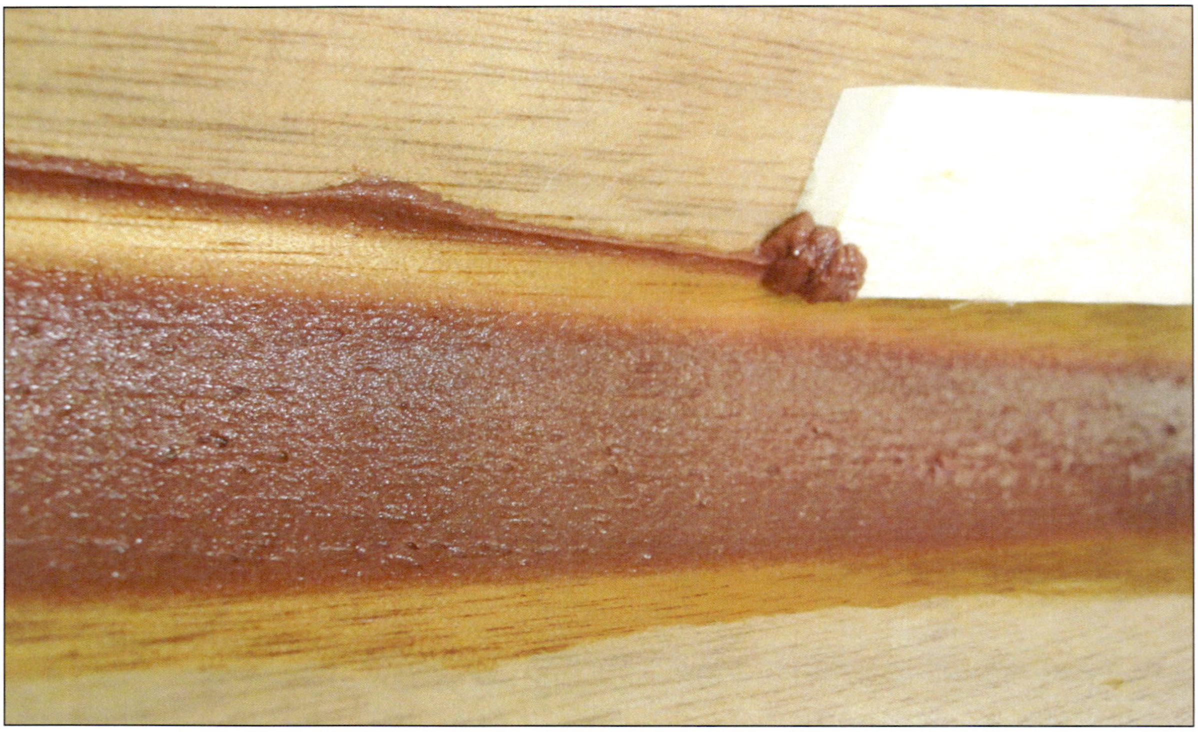

Filleting inside corners is difficult. Extra epoxy should be added in the corner itself and the fillets formed from the corner outward.

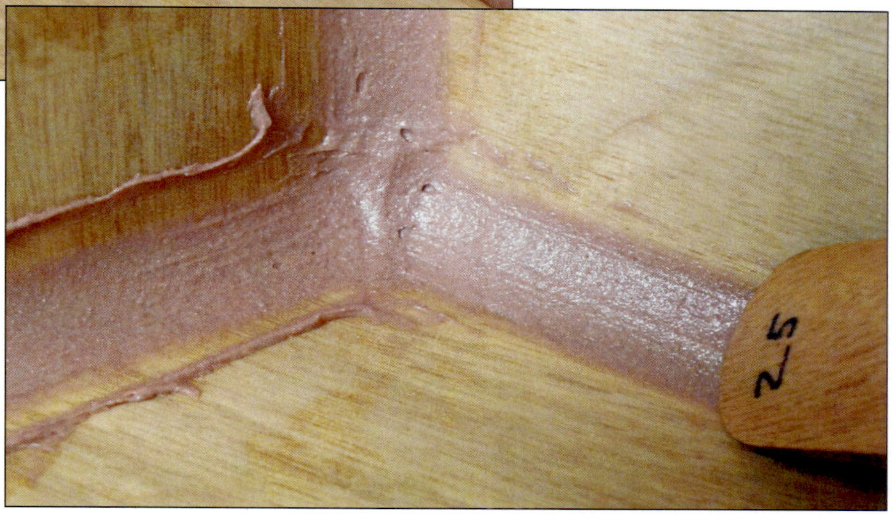

A far easier way to fillet inside corners, is to fillet one edge and allow that to cure before filleting the other edges.

The photos below show the PT 11 main bulkhead and gussets. These fillets are rather complex, very structural, and a good example for more advanced filleting techniques.

Here we chose to fillet the bulkhead edge first (photo below) and the gussets later. The photos of the gussets being filleted (following pages) show how the inside corners can be formed much better if one edge has already been filleted and allowed to cure before filleting the other edges.

The little imperfections in the middle of each fillet are what we call "swooses". They are the unavoidable result of having to form the fillet from the ends toward the middle (see next photo) and they are easy to remove later with sandpaper wrapped around a piece of dowel.

Forming the fillet from the ends toward the middle will leave gaps at both ends (right side of photo) that will be filled when filleting the other edges.

To avoid having to sand, we filleted the gussets later the same day, after the bulkhead fillets had cured enough to be firm. For this to work though, the residue at the edges of the fillets (at least near the gussets) must be cleaned up carefully with a sharp chisel stick.

As discussed before, we always prime for fillets with un-thickened epoxy. It makes fillets adhere better and it makes filleting easier, but it should be applied very thin, or the excess wiped up with paper towel.

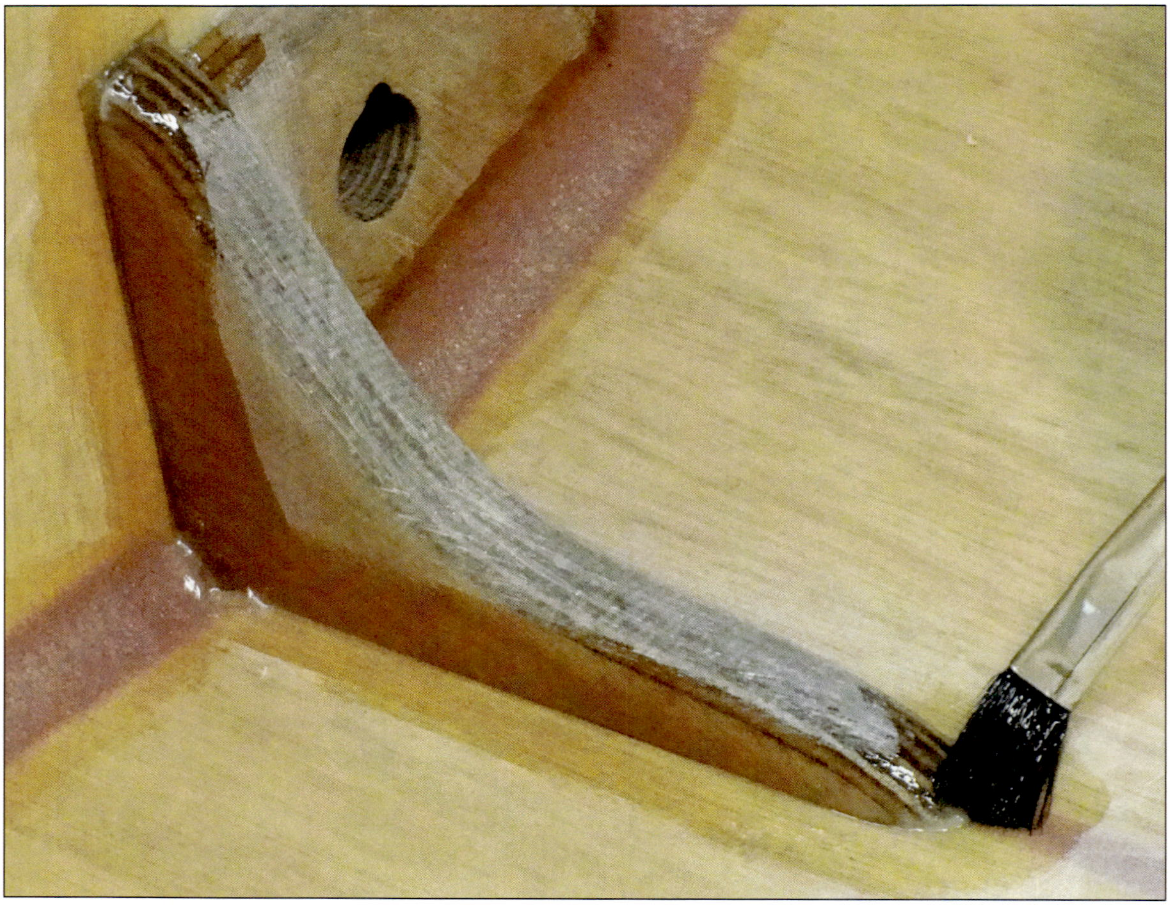

The gusset fillets are formed by starting in the middle of the inside corner and filleting toward the gusset tips.
Laying the stick back at a flatter angle will increase the size of the fillets at the tips.

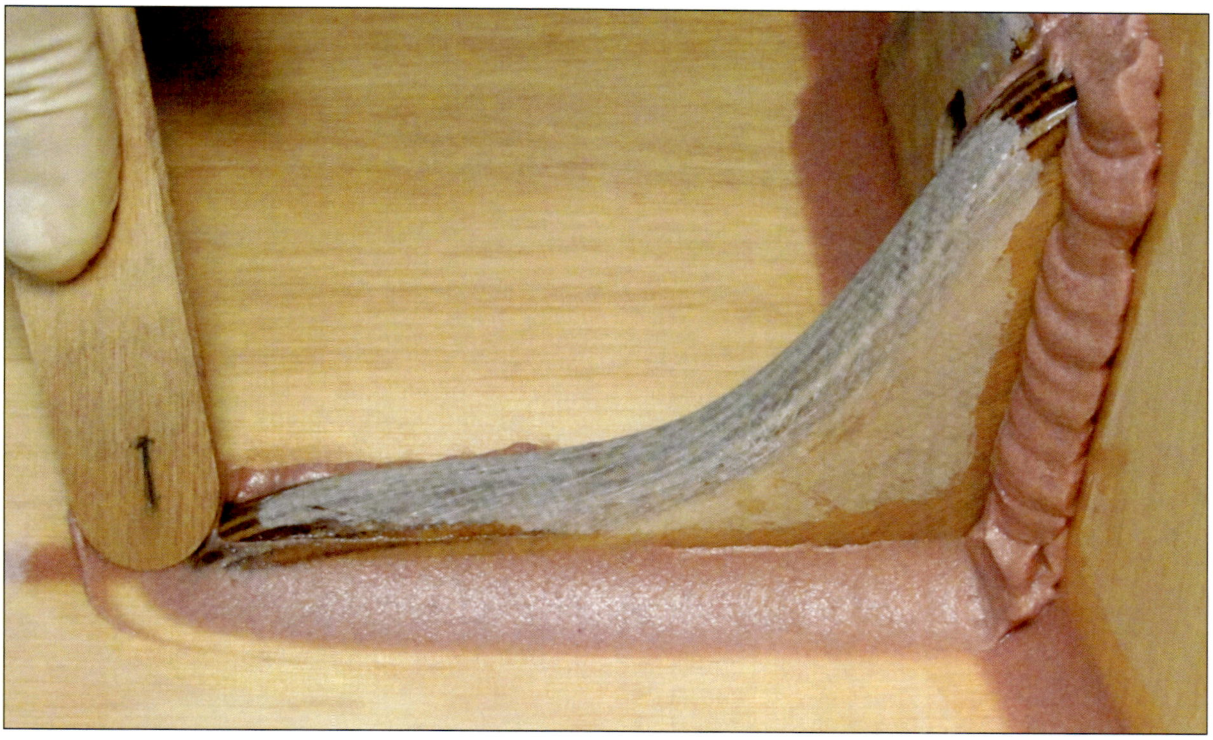

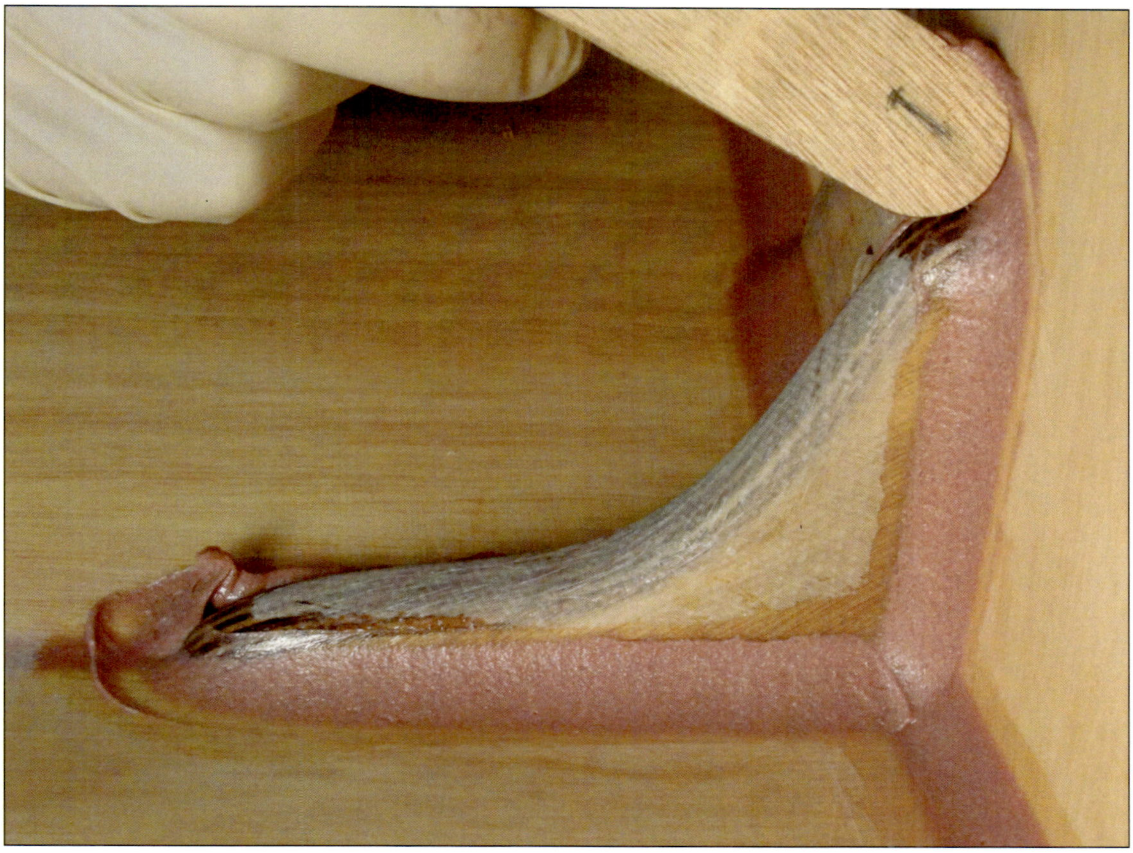

The inside corners look pretty good, don't they? That's good, because inside corners are really hard to sand.
Separating filleting operations, even by just a few hours, makes these kind of intersections much cleaner.

For us, reducing the amount of time spent sanding is worth the time spent preparing and cleaning up after.

Sharp chisel sticks save a huge amount of sanding time. It takes a careful touch to clean up around fillets without gouging into them, but again it's worth it.

SANDING FILLETS

...is easier, if done within a couple of days of applying them.
Start sanding by wrapping a piece of sandpaper around a dowel to sand the fillet itself and then use hand sandpaper to sand the edges and corners.
Avoid cutting deeply into glass or coating (on pre-finished panels) when sanding fillets.

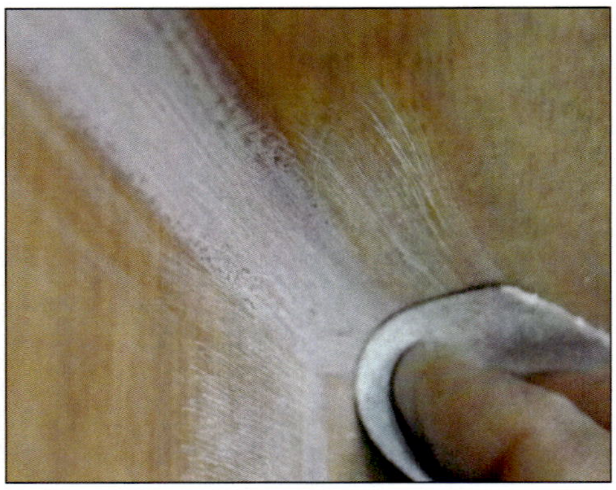

TROUBLE SHOOTING FILLETS

If you have trouble with fillets, here are a few tricks to try:

- Mix the epoxy a bit thicker. The mix should be thick enough for a large blob to hang off the end of the stir stick without drooping.
- Adding more 406 Colloidal Silica to the mix will make it more workable.

- If a fillet starts to slump or is otherwise bad, form it with a smaller stick and form the final fillet later, after the first fillet has cured, or started to cure.

- If you have a void or bubble in an otherwise good fillet, leave it to be filled later.

- Clean the tip of the fillet stick between passes.

The important thing is to clean up well with the chisel stick before leaving the scene. **Remove all the residue left on either side of the fillets while the epoxy is wet, not later when it's hard.**

GLASSING

Laminating light fiberglass cloth on plywood can do great things for the strength and longevity of plywood. Glassing both sides of plywood (especially thinner plywood) makes a marked increase in stiffness and strength because it's always the outermost fibers (on any beam or panel) that take load first when bending.
Strength aside, plywood surfaces that are exposed to chafe, water, and sun should be glassed unless the boat is used lightly and kept in a garage.
Glassing is work, yes, but it can be fun and it allows the possibility of a very long-life structure.

The dry method of glassing is used for most of the parts on our boats, where the fiberglass cloth is positioned on the surface dry, and then wet from the top down with a roller or squeegee as shown in the photo below.

**Roller covers, roller pan vs palette on page 47.*

SATURATING OR "WETTING OUT" FIBERGLASS CLOTH WITH EPOXY

When saturating fiberglass cloth, a squeegee can be helpful for quickly moving a large puddle of epoxy over a big area, but the roller is an important tool for evenly wetting out fiberglass cloth.

The roller can be used exclusively on all the parts pre-glassed on the bench because using the squeegee on these parts can result in resin running down the edges of the parts and gluing the parts to the table, etc.

When using a roller to saturate fiberglass, **the trick is to have the roller very wet** (evenly wet, see photo and text on page 34), put the roller down on the dry cloth and roll only a short distance (8 or 10 inches). This "anchors" the cloth to the surface and applies enough epoxy to saturate the cloth. Repeat the wetting and rolling, rolling away from the anchored area to avoid wrinkling the cloth.

For surfaces that will be clear finished (not painted), use plenty of epoxy and avoid pushing a dry-ish roller onto dry cloth.

When the surface is completely 'wet out', finish by rolling over the whole surface (firm pressure on the roller) to achieve an even, textured surface.

Look for an even texture with no glossy wet spots. Overly wet areas can be remedied by wrapping a clean rag around the roller and squeezing. This takes resin out of the roller and it can then be used to roll over the wet areas, soaking up excess epoxy.

Many of the parts of our kit boats are pre-glassed on the bench before assembly.
The outside of the hull is glassed after assembly, but the inside surfaces of the hull panels and all other exposed surfaces are glassed before assembly, while the parts are laid flat on the table.

Glassing, fill coating, and sanding all these exposed surfaces before assembly is a technique that we take advantage of wherever possible.

In this photo, parts for a dinghy have been 'nested' together and covered with light cloth. The cloth was saturated with a roller and later in the day given a "fill coat". (see next page)

The exposed surfaces of these hull panels are being pre-glassed on the bench. The areas that are not exposed (inside the back and forward seats) **are not glassed, but are given a first coat of epoxy when saturating the cloth** (photo on left) **and given a second coat when fill coating the glass** (photo below).

FILL COAT

The fill coat fills the weave of the glass cloth to an almost glossy surface, so that you can sand the surface without cutting into the glass fibers (see above photo).

The fill coat must be applied the same day as the fiberglass is applied, and here's why. The textured surface mentioned on page 24, cannot be sanded (to prepare for the fill coat) without cutting into the fibers. **Cutting into the fibers of the cloth, even a little bit, destroys the strength of the fiberglass, so the fill coat must be applied when it can "chemically" bond.**

Achieving a "chemical" bond with the fill coat is especially important when glassing panels that will have structure bonded to them later. Pretty much all glassed parts and panels have something bonded to them somewhere, mostly on their edges, so making sure the fill coat sticks to the fiberglass is important. Because this must happen the same day, the glassing has to be done early enough to avoid fill coating in the middle of the night.
How hard does the epoxy need to be before you can fill coat? Sticky is not quite hard enough, but rubbery is fine. Try pulling up on the edge of the cloth, if it resists, you can fill coat.

Apply the fill coat with a roller. This is much easier than saturating the cloth, and takes about two thirds as much epoxy.
Look for an even, glossy surface with only a slight texture showing as in the above photo.

WEST SYSTEM 410 Microlight® Fairing Filler can be added to the epoxy for fill coating surfaces that will be painted. This makes the final fairing easier and it also allows the fill coat to be applied thicker. Using about 25 percent by volume filler to epoxy will allow the thickened fill coat to be applied with a foam roller.
For a really thick fill coat, the microlight filler can be mixed in at almost 1 to 1 by volume, but it must be applied with a nap roller. **Here's how:** Mix your batch of epoxy in something tall, like a paper coffee cup, pack the microlight into a similar cup to nearly the same level, mix the two together in a bigger cup. **Apply with a short (¼") nap roller.**
These rollers soak up a lot of epoxy, so maybe cut a 7" or 9" roller in half.
Get the batch of thick epoxy onto the part and then roll it out evenly. The thick fill coat will be quite bumpy, so follow with a foam roller to flatten the surface.

MORE NOTES ON GLASSING

Most glassing on our boats is done over bare plywood. Random orbit and hand sanding plywood with fine paper before glassing will help the cloth lay flatter. **Sanding out all scratches in wood with fine sandpaper is necessary for surfaces that will be bright finished.**

Bare wood can soak up epoxy, especially on edges (end grain), so look for dry spots in these areas, and roll on more epoxy where needed.

The secret to happy glassing is to be prepared and work fast. Working fast is important. You want to be finished before the epoxy starts to thicken.
On big areas, think big picture.

Make long passes with the roller and use enough epoxy. You can always transfer some from an overly wet area to a dry area with the roller after the whole area is "wet out".

Wrinkles can often be "pulled out" by tugging on the edges of the cloth or using the roller: **The photo on the right shows pulling out that wrinkle on the chine by angling the roller handle away from the wrinkle while rolling straight ahead** (arrow).

Use the roller for inside and outside corners wherever possible (photos above).

The **"wet method"** is sometimes used on vertical surfaces such as this transom.

The transom has had epoxy rolled over its surface before applying the cloth.
This method works best when the cloth can be applied without wrinkles, usually by rolling the cloth out over the wet surface as shown.

SANDING GLASSED SURFACES

Glassed and fill coated surfaces should be washed to remove amine blush if needed (See page 40) and **sanded with a random orbit sander** with 100 or 120 grit sandpaper until only tiny glossy spots remain.
If a fair surface is desired, some block sanding will be needed.

Block sanding before power sanding will remove the high spots and flatten overlaps that the random orbit would just ride over without flattening.

Do not sand deeply into fiberglass cloth, especially in areas where the glass is holding the boat together, (chine tapes are one such area). If the weave of the cloth is visible when sanding, it means the fill coat was not adequately thick or the sander had too much coffee. Either way, sanding into the weave of the cloth destroys the strength.

CUTTING FIBERGLASS CLOTH

When cutting from a roll of cloth, **pulling a strand and cutting along the weave** is the only accurate and clean way to cut cloth. It will prevent a mess of strands coming out of the edges of the cloth and will promote clean glass work.

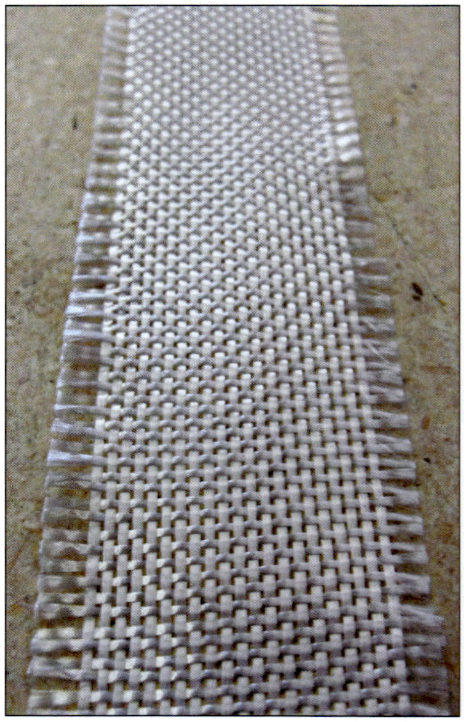

This method is used when cutting "tapes" (narrow strips) used for chines and edges.

Cutting your own tapes is work, but will provide much better results than using commercially made tapes that have a thick selvaged edge. **Tapes cut from a roll of cloth naturally taper on the edges as a result of pulling loose strands from the edges.**

Tapes are usually cut from the end of the roll of cloth (not the edges) and overlapped where long lengths are needed.

Cutting can be done with sharp scissors or a snap-off blade knife.
- Start by making a deep slit on the edge at the desired width.

- Pick a single strand from the edge of the slit.

- Pull on that strand while reaching out with the other hand and holding the cloth to the table. Reach farther and farther while pulling on the strand to make a visible line all the way across.
Pulling the strand completely out will make for an easier line to follow, but is hard to do with some types of cloth.

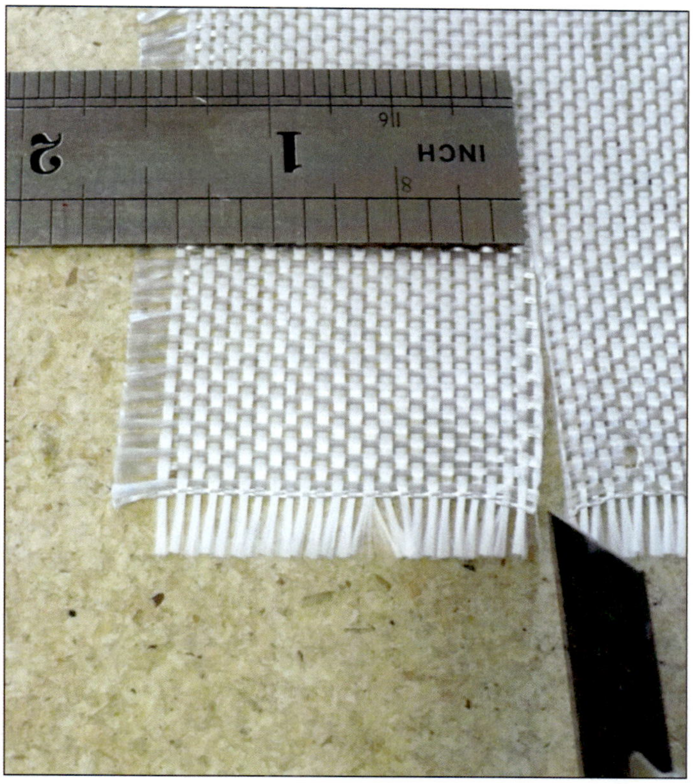

- Pull on the edges to remove the "puckers" before **cutting along the line**.
Sometimes the line can be hard to see. Moving your vantage point can help.
Once the tape is cut, pull a couple of strands from the edges of the tape and set it carefully aside.

GLASSING EDGES & CORNERS

Edges are vulnerable areas in wood/epoxy construction.
Because end grain plywood absorbs water easier than elsewhere, a small nick through the glass on an edge will cause fiberglass "zippers", where swelling wood can cause the fiberglass to fail.

The treatment of corners and edges could merit a whole book, but we'll cover some of the methods that we use.

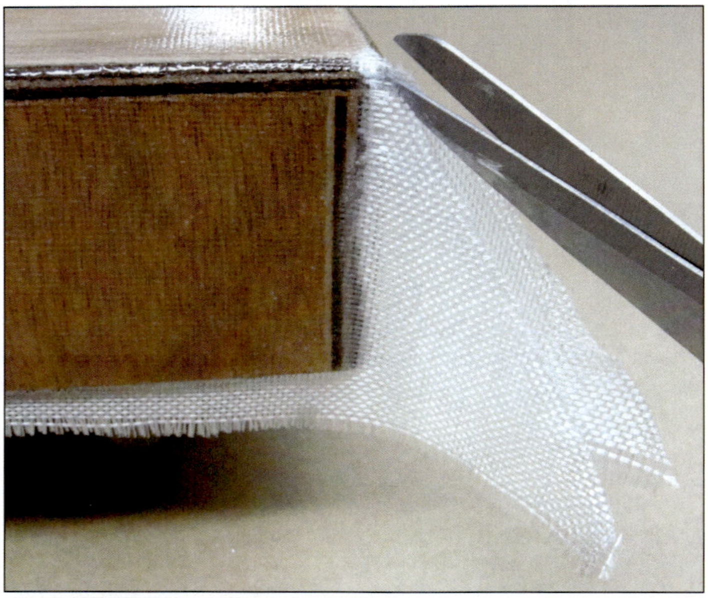

Fiberglass can conform to some crazy shapes (see next page), but it *can't* wrap over sharp edges or corners.
If you are using a router to round the edges, a 3/16" round over bit will make about the smallest radius that glass can wrap around. A 1/4" radius (shown) will make glassing easier.
The outside corners of this hatch lid are shown being glassed.
(1) The top surface and edges are wet out first.
(2) Sharp scissors are used to cut down the middle of the fold, cutting all the way to the top surface.
(3) One edge is held out of the way while folding the other edge around the corner with a wet roller.
(4) The second edge is wrapped around the corner with the roller.

This example has rounded corners.
The trick with getting glass to conform to edges like this (without cutting and overlapping) is to cut away all excess cloth before wetting the cloth (second photo).

The reason the cloth is able to conform to these corners is that the fibers *at the corner* are on the bias.

Experimentation with scrap cloth will show that fiber orientation will make glassing a complex shape either possible, or impossible.

A small rectangle of cloth with the fibers running parallel to the edges will not conform well, (try wrapping it around your fist), but the same size rectangle with the fibers on the bias will easily wrap around just about anything, as long as it doesn't have sharp edges.

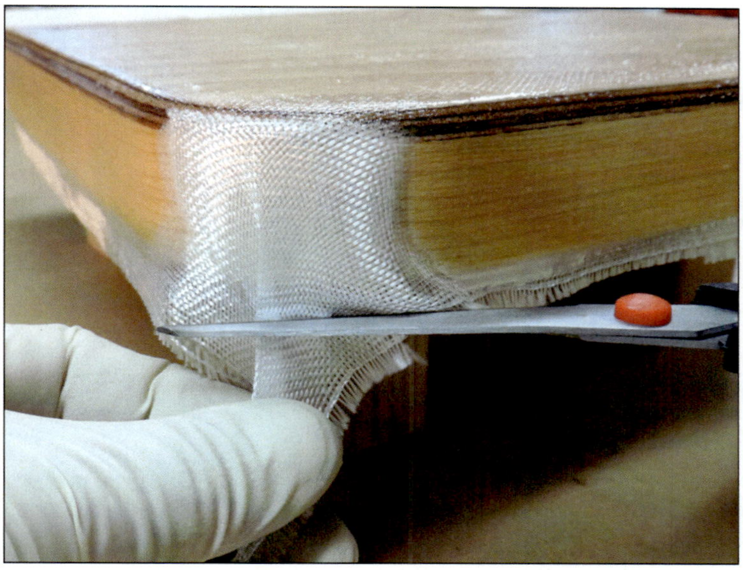

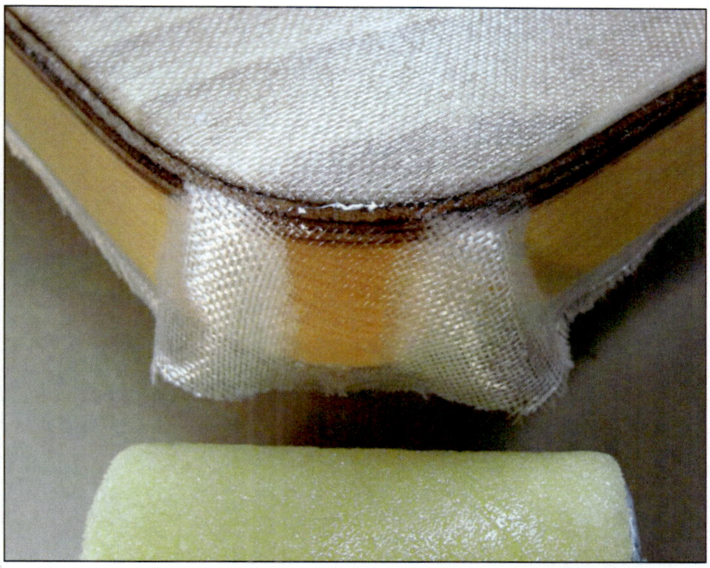

Often it's not possible to glass both sides of something at one time. Glassing one side at a time means the glass overlaps on the edges, but the cloth must be carefully trimmed with the knife and sanding block *on a rounded edge*.
The router (same round over bit, same bit depth) will trim the glass nicely too.

Tip: If the glass tries to lift away from an edge, thicken a tiny bit of epoxy with 406 Colloidal Silica® (make it thick), lift the cloth and rub a small amount on the edge (with a gloved fingertip) and put the cloth back down. This works wonders, but can cloud the transparency.

On some edges it's not practical to round a plywood edge and wrap glass around it, so here we "cap" the edges with multiple layers of cloth, then trim them flush with a knife and sanding block.
This method, cutting narrow strips of cloth and laminating them onto the flat edge with a roller, makes a pretty tough edge, but the finished edges must stay fairly sharp as illustrated below.

The more layers used, the tougher the edge... and also the harder to trim.
Warm with a heat gun (and trim the following morning) to make trimming easier

G-FLEX® epoxy from WEST SYSTEM® offers alternative methods for toughening edges.
We haven't used this product much in this way, but have read that it can be applied to areas like oar and paddle blade edges (without glass) and the result is an edge that can be whacked hard enough to dent the wood without damaging the epoxy coating.

Our two-piece dinghies must have sharp and very tough hull edges at the joint between the hull-halves. **We chamfer the sharp hull edges with a sanding block and then laminate two layers of masking tape (to make it stiffer) and apply it as a dam for injecting with epoxy.**

We fill these dammed chamfers and block sand flat before glassing the hull and bulkheads, so that <u>**the glass on both faces covers the filled corner**</u> (photo's at bottom).

As you can see, this technique can be used on complex corners. We use high density fillers to thicken the epoxy which makes block sanding flat a bit of work, but the finished corner after glassing is very tough.

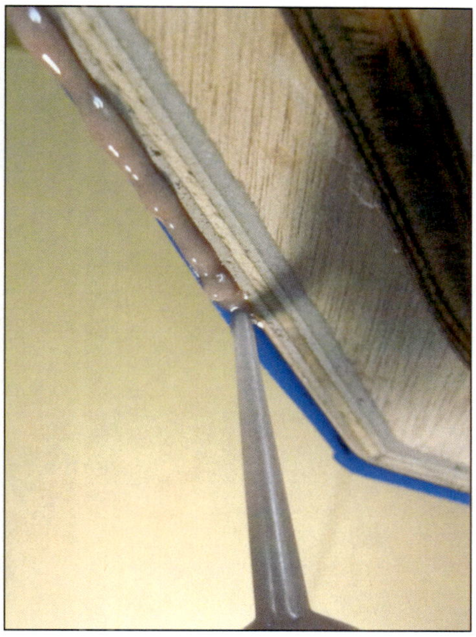

Both faces are glassed (separately), capping the filled corner.

COATING

This section applies to bare wood and pre-glassed surfaces.
On bare wood, three coats of epoxy are recommended for chafe resistance and as a moisture barrier, but the second coat can look so good that often the third coat is skipped.

Pre-glassed and fill coated surfaces generally require only one finish coat.

The final coat of epoxy on either surface we call, the "gloss coat".

Applying epoxy for coating is best done with a yellow foam roller, except inside corners, where a throwaway bristle brush is needed. *(see page 37)*
Cut the roller to the length needed for the job, this will save epoxy and rollers.

Wetting the roller evenly is important and a pallet works best for smaller jobs.

Pour a puddle of epoxy (as shown) in the middle of the pallet, **set the roller down on the edge of the puddle and pull toward you to create a wet path of epoxy.** Roll over the wet area a couple of times until the roller is evenly wet.

Epoxy should be metered onto a surface by wetting the roller, from a pallet or roller pan and quickly rolling it out over a small area and repeating the wetting and rolling until the whole surface is roughly covered.

Then, roll in one direction and then the other multiple times to make an even film of epoxy.

Full length passes with firm pressure on the roller are best for making an even coat.

On bare wood, the first coat of epoxy soaks in and lifts the fibers so that the surface is rough. **The first coat needs to be thorough, but it doesn't need to be thick**. Just make sure the whole surface is wet.

When cured, this surface **may** need to be washed to remove *amine blush*, but it definitely needs to be sanded. *(See page 40)*

Sand until relatively smooth with 120 grit, or so, but don't sand through into the wood. On corners and edges, sand with finer sandpaper to avoid removing too much epoxy.
You don't need to remove every last tiny glossy spot before the following coat.

The second coat of epoxy should be heavy enough to provide some protection, but not so heavy that drips, sags, runs, or curtains appear.

Drips and runs look horrible, and sanding out drips results in a thinning of the nearby coating, so meter out the epoxy evenly and roll over the surfaces repeatedly before tipping.

The second coat needs to be "tipped" to remove the 'stipple' (photo on right) left by the roller. Tipping is best done with a wide foam brush (see photos on following pages), making full length passes when possible.
Epoxy will "flow out" best if the tipping is done as soon as possible.

Smaller areas can be tipped more economically by splitting a roller cover into thirds, (bandsaw, or pull saw) and using as a brush.
A handle with a narrow slot to hold the piece of roller can be made from wood.

GLOSS COAT

The final coat of epoxy is what we call the gloss coat, whether it's a finished surface on the inside of a boat or a final coat before painting or varnishing.
Surfaces exposed to the sun should be given a UV resistant coating, such as paint or a good varnish, but the epoxy gloss coat can look quite good.

The gloss, or final coat in most places happens after the boat is assembled.
On the inside of a boat, the challenges are metering the resin evenly over these uneven surfaces (with a roller and brush), and getting the area in question tipped while the resin is still thin enough to 'flow out'.

WEST SYSTEM® 105 RESIN with 207 Special Clear Hardener® is a really good epoxy for coating, especially for the gloss coat. It flows out beautifully and is totally clear. It has a decent amount of working time and does not produce amine blush.

Removing dust from surfaces before gloss coating is a challenge.

Tack rags are a must, but most of the dust should be gone before using the tack rag.

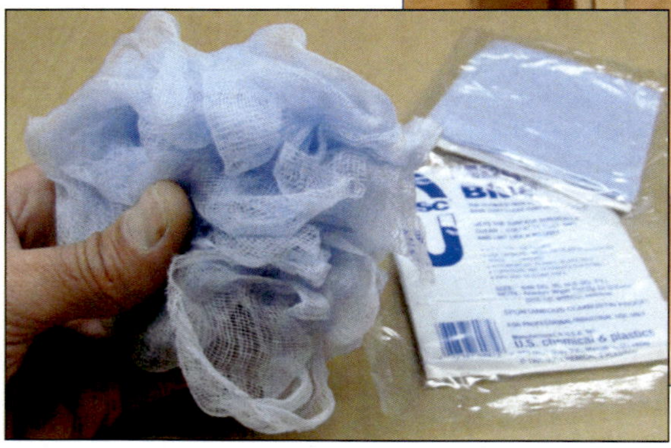

Start by vacuuming with a small brush attachment and then wipe the surfaces with clean cotton rags (or the blue, lint-free paper towels) until the rags are not picking up any more dust.

Unfold and lightly bunch the **tack rag** and glide over the surface without pushing hard or catching on edges.

If a tack rag shows more than a tiny amount of dust on its surface, its no longer picking up dust, so re-fold and bunch the rag or replace it.

Apply epoxy to inside corners with a throwaway bristle brush.
Use less epoxy when brushing than you would think, or you will get runs from the corners.

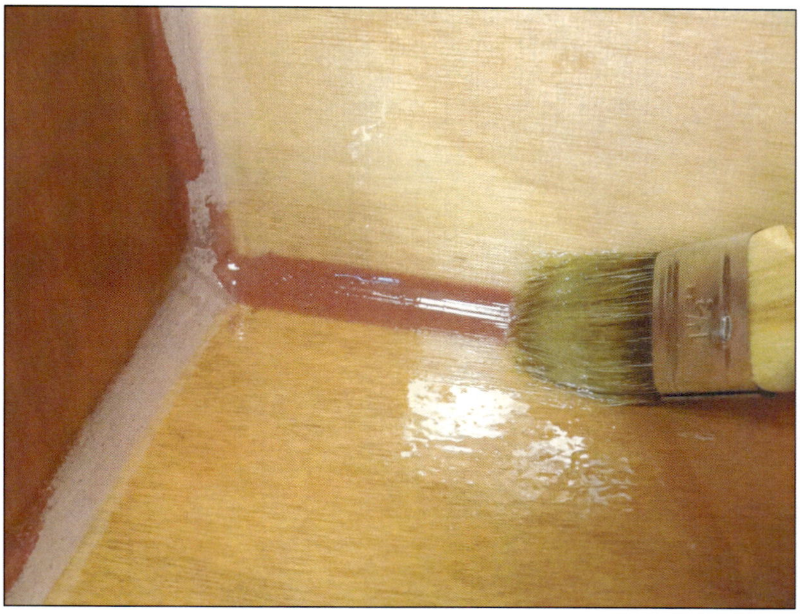

Use a bristle brush to apply epoxy on other complex areas, but use the roller wherever possible.

We give throw-away style bristle brushes a haircut with scissors, cutting off a bit of length and cutting the width in half. They work better and drop half as many bristles.

Brush cleaning page:48

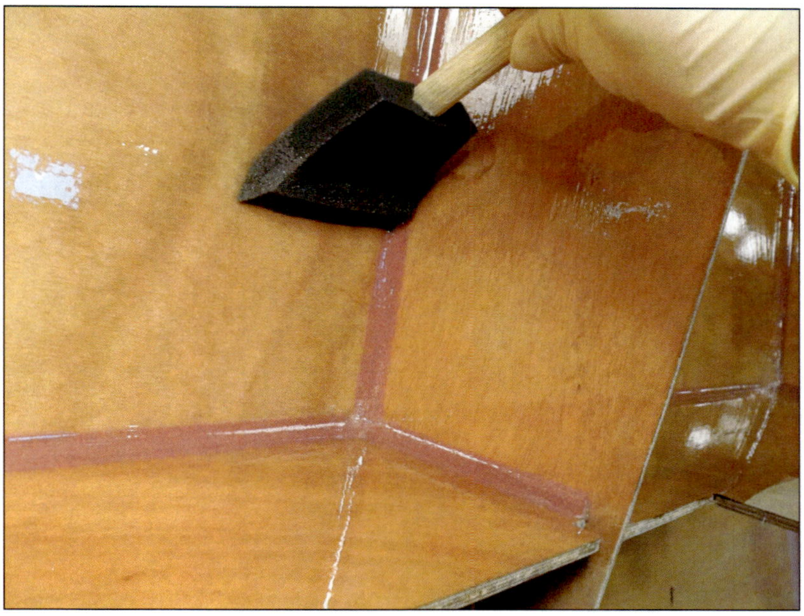

Gloss coating can be broken into sections (by taping) to keep it from being too daunting.

Inside corners (the middle of a fillet) are a good place to stop and start. One could separate all the seats from the hull sides, etc.

One can prepare surfaces next to the tape with a piece of 3M® "scrubbie" (Scotch-Brite® scouring pad 07447). *See page 46*

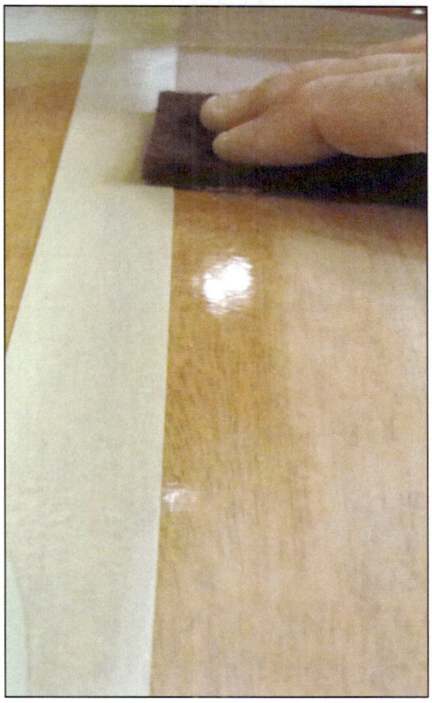

PRIMER?

We rarely use primer before painting. Instead, we gloss coat surfaces to be painted.

Using an epoxy gloss coat instead of primer has some advantages:

Epoxy "flows out" and primer doesn't. This means less sanding before painting, but more importantly, the epoxy is glossy (as the paint will be) allowing one to see imperfections that need to be filled or faired before painting. Epoxy is also tougher and a much better moisture barrier. Most paints adhere well to fully cured and sanded epoxy.

Primer is sometimes used as a fairing compound, but here the surfaces must be fair before gloss coating. See about adding 410 Microlight® Fairing Filler (pg 26) to the fill coat and "Abrasives, Sanding & Fairing" on page 45.

On large jobs, a short roller extension can help apply more pressure on the roller and ease the pressure on your wrist (below).

We almost always tip the epoxy with a wide foam brush after rolling it out evenly.

Temperature greatly affects the viscosity of epoxy. A bit of heat from a heat gun or hair dryer can be used when gloss coating to make the epoxy 'flow out' better.
Hold the heat gun in one hand and the brush in the other. Aim the heat just ahead of the brush. This works great, but **too much heat is a quick way to ruin a job.**
Never hold the tip of the heat gun close to the surface and always keep it moving.

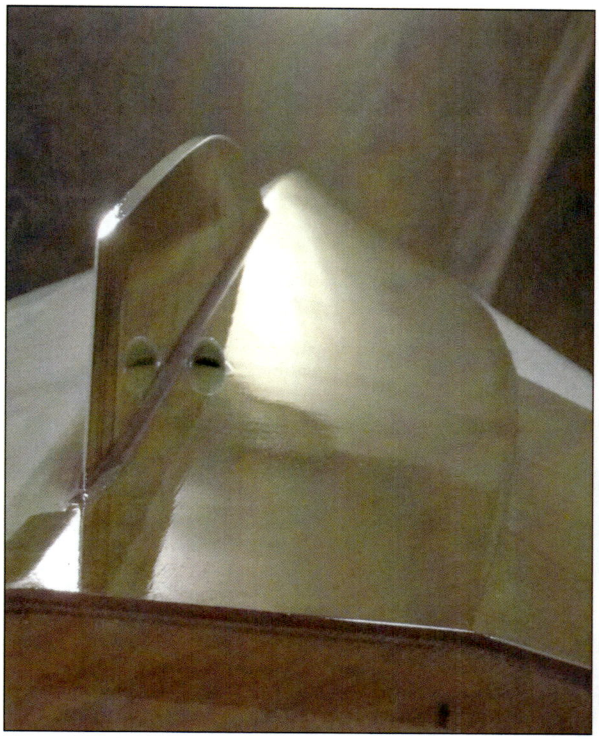

Small parts are often coated separately before installation, which can be great if they are supported in a way that allows access for the roller & brush.
We use hot-glued blocks of wood or nails or screws attached to a base of scrap plywood.

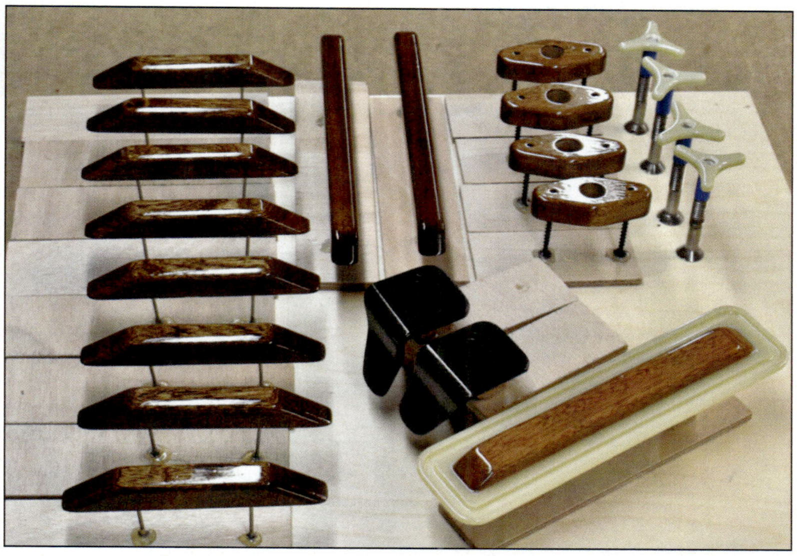

AMINE BLUSH

This is a waxy film that rises to the surface of epoxy when using WEST SYSTEM® 205 Fast Hardener® & 206 Slow Hardener® and other epoxies. Amine blush is not an issue with 207 Special Clear Hardener®.

Amine blush is water soluble, and is easily removed after the epoxy cures with a wet rag or "scrubbie". (3M® Scotch-Brite® scouring pad 07447)

Warm water is good if you have it. Use a small bucket and repeatedly rinse the rag in the bucket.
Water can stain raw wood and often amine blush must be removed on areas that have raw wood nearby. Lots of pressure on a repeatedly rinsed damp rag will remove the amine blush without getting water everywhere. The epoxy can then be sanded.

If the area you are working on can get totally wet, and if you want thorough preparation without lots of sanding, use a piece of 3M® "scrubbie" and water.
This is a good technique on areas like the outside of the hull between coats of epoxy.

Amine blush will be more evident in damp shops.

Some of the best epoxies produce amine blush, but it must be removed before sanding.

SURFACE PREPARATION

Many structural joints rely on "secondary bonding". This means that joints occur between panels and parts that have already been "glassed" or coated with epoxy.
It is a great advantage to be able to pre-glass or pre-coat parts before assembly, so an understanding of surface preparation is important.

The best surface for secondary bonding is one that has been well sanded with fresh, sharp sandpaper.

Before sanding, these surfaces can be washed to remove amine blush, (if necessary) or wiped with acetone to remove contaminates (if necessary).
After sanding, the surfaces should be vacuumed with a clean brush attachment, and wiped hard with a clean, dry, cotton rag to remove fine dust left on the surface.
Mechanic-style, lint-free paper towels (often blue) seem to work well for de-dusting. They both pick up and show dust and seem to be lint-free.

Inside corners and rough surfaces can be hard to sand to a gloss free surface when prepping for re-coating or paint.
Use pieces of 3M® "scrubbie" (Scotch-Brite® scouring pad 07447) to remove gloss from those areas. Wrapping a piece of "scrubbie" around the end of a stick works well for hard to reach areas.

Thinners are not a good idea for surface preparation, unless wiping to remove contaminates before sanding. All hardware store thinners leave a residue that can inhibit a secondary bond and often there are compounds in rags or paper towels that are broken down by thinners and left on the surface.

PEEL PLY

Peel ply is light nylon cloth that can be applied over fiberglass tapes, (chine tapes, seat edges, etc.)

It is removed after the epoxy cures.

Using peel ply can dramatically reduce sanding time and eliminate the need to fill coat.

This photo shows two identical fiberglass tapes. **The one on the left was peel ply'd and is as smooth as it looks.**

Removing peel ply after the epoxy has cured.

Extra epoxy must be added to the fiberglass when using peel ply for two reasons:

One; the fill coat *(see page 26)* happens now, not later.
Two; the peel ply absorbs a bit of epoxy.

If the cloth is not "wet" enough with epoxy before applying the peel ply, pesky "pinholes" will show themselves later when finishing.
Rolling or brushing on extra epoxy before applying the peel ply (as shown in the photo on the right) will make for a smooth finished surface and edges.

Extra epoxy added to the glass before applying the peel ply.

Peel ply strips should be cut much wider than the tapes, and **pressed down into the wet fiberglass** with a folded rag. Any extra epoxy will come up through the peel ply and into the rag.
Pieces of folded towel or paper towel work best for applying even pressure.

If un-treated peel ply is used, the surface left after removing it should be contaminate-free and ready to bond to. The surface left after removing the peel ply will be textured and will require a small amount of sanding before finish coating.

Peel ply will not conform like fiberglass cloth can, so complex corners require multiple small pieces of peel ply.

Different weights of peel ply are available, the heavier peel ply absorbs more epoxy, but will leave a flatter surface. Lighter peel ply conforms better to complex areas.

Cutting peel ply strips on the bias (diagonally from the roll) can be an advantage because the strips conform to complex shapes better and because there will be no pesky nylon strands left in the cured epoxy from the edges of the peel ply.

Large areas can be peel ply'd (to avoid having to fill-coat), but it can be challenging to apply large pieces without getting wrinkles in the peel ply.

FILLING HOLES WITH A SYRINGE

Some people try to avoid drilling lots of holes when building a boat. We drill as many holes as we need to and don't worry about it at all, because filling holes with epoxy in a syringe is so easy.

Unlike a putty knife, filling holes with a syringe fills the holes all the way, (no coming back later after sanding to fill again). Most importantly, the mounds left by the syringe are a fraction of the effort to remove (with a sanding block) compared to the smears left by a putty knife.

To **color match** (for clear finishes), try thickening the epoxy with colloidal silica and add small amounts of 407® (brown) and 410® (tan) fillers to darken for a color similar to the area where you are filling holes.

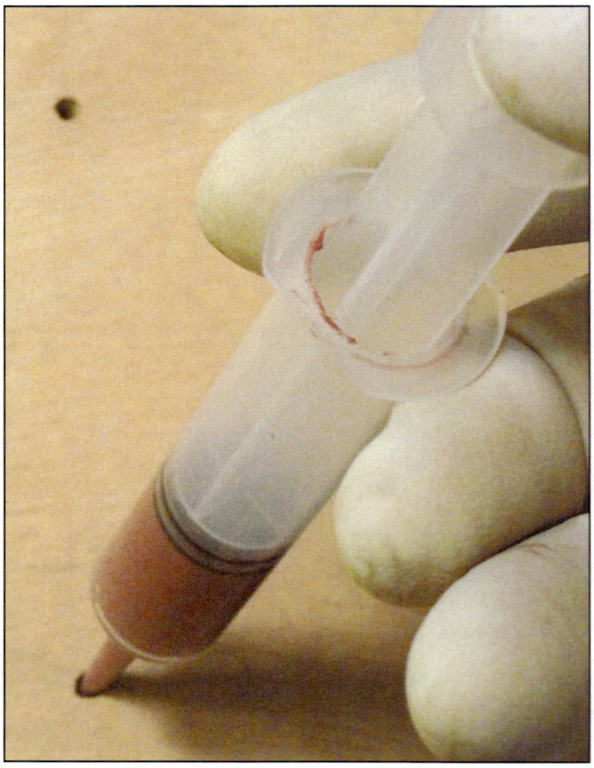

Epoxy can be packed into the syringe with the stir stick as shown.

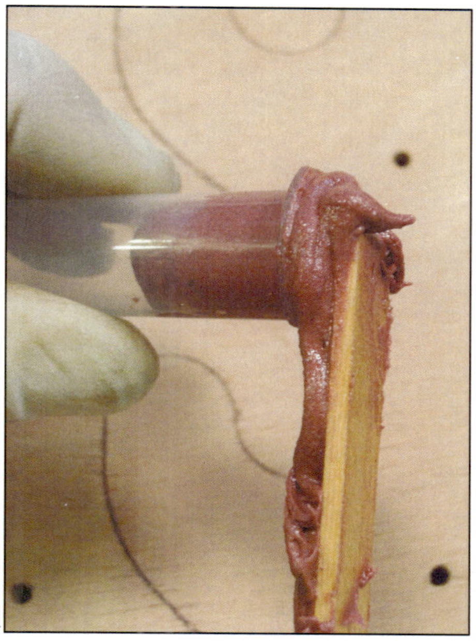

To get multiple uses out of a syringe, either clean with acetone or leave a tiny bit of epoxy in the syringe when finished. After this cures hard, cut a small amount of the tip off, flex the tip and pull the plunger out. The rubber part will have epoxy stuck to it, remove it and bash it with a hammer, then pick off the epoxy.
If that doesn't work and the plunger is too tight to use the syringe, fold sandpaper in a U-shape and sand the plunger to fit.
A new syringe is needed for small holes and good flow control, but sometimes a crusty old syringe can still do the job!

SELECTIVE FILLING

Part of getting a surface fair is applying thickened epoxy where needed **before you start fairing.**
My preference is to use tools to carefully apply the thickened epoxy where it is needed instead of gobbing it over the whole surface just to sand most of it off later.

The filler shown being applied below fairs an overlapped layer of heavy fiberglass.
The thickened epoxy, a mixture of 410 Microlight® and 406® silica is applied as shown on the left and then "tooled" to a flat plane as shown on the right.
Note the pressure on the wide putty knife being applied to the edges to avoid bending the blade.

This kind of filling often requires a couple of tries before a final pass.

Tools for selective filling can include different width putty knives, drywall knives from the hardware store (the wider blades below), straightedges, and even custom, extra wide and stiff tools like the one shown at the bottom of the page.

What do they all have in common? The blades are sharp and the edges are kept clean.

A blade can be stiffened by double-side taping something stiff to one face of the blade, or made more flexible if needed by removing the handle & frame (upper right).

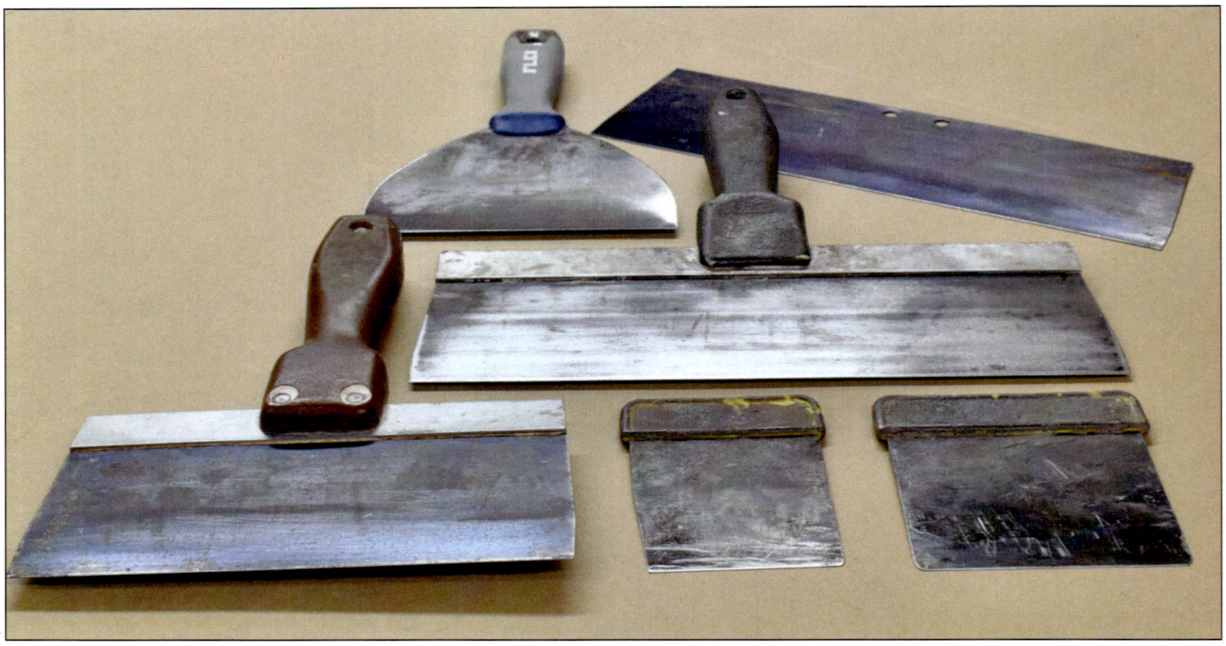

Big fills need bigger tools. If a flat surface is desired, a stiff tool is needed. Just remember that the working edge of the tool should be sharp and straight.
Horizontal fills (shown below) can use epoxy that isn't mixed quite as thick, which is helpful as the excess rolls under the tool and applies itself where needed.

ABRASIVES, SANDING, & FAIRING

Why do we use sanding blocks & longboards? Because it is the only way to get from a bumpy, unfair surface to a flat or fair surface. They sand away the high areas only, until a fair or flat surface is achieved.

For a flat surface, a thick and stiff sanding block will work best. **For curved surfaces,** a sanding block or longboard that will bend to the surface being faired is necessary (below right).

On the right are two longboards, one thinner and more flexible than the other. These plywood longboards, about 18" long are used for fairing larger surfaces.
The shorter plywood sanding blocks of different thickness & stiffness are used for smaller surfaces.

Both the longboards and sanding blocks can use the adhesive backed roll sandpaper shown on the left of the photo. This roll sandpaper is about 2 3/4" wide, so the blocks and longboards are made to that width. Sanding blocks without handles have sandpaper on both sides.

There are many levels of quality in sandpaper. Changing your sandpaper often will produce a better job with less effort.

Custom tools for sanding can often produce the desired results, but they take some imagination and effort.

The random orbit sander is used often, but it will not make a surface flat or fair.
A combination of longboard or block sanding to remove the high spots and random orbit sanding to finish can work well and is far faster than doing it all by hand.

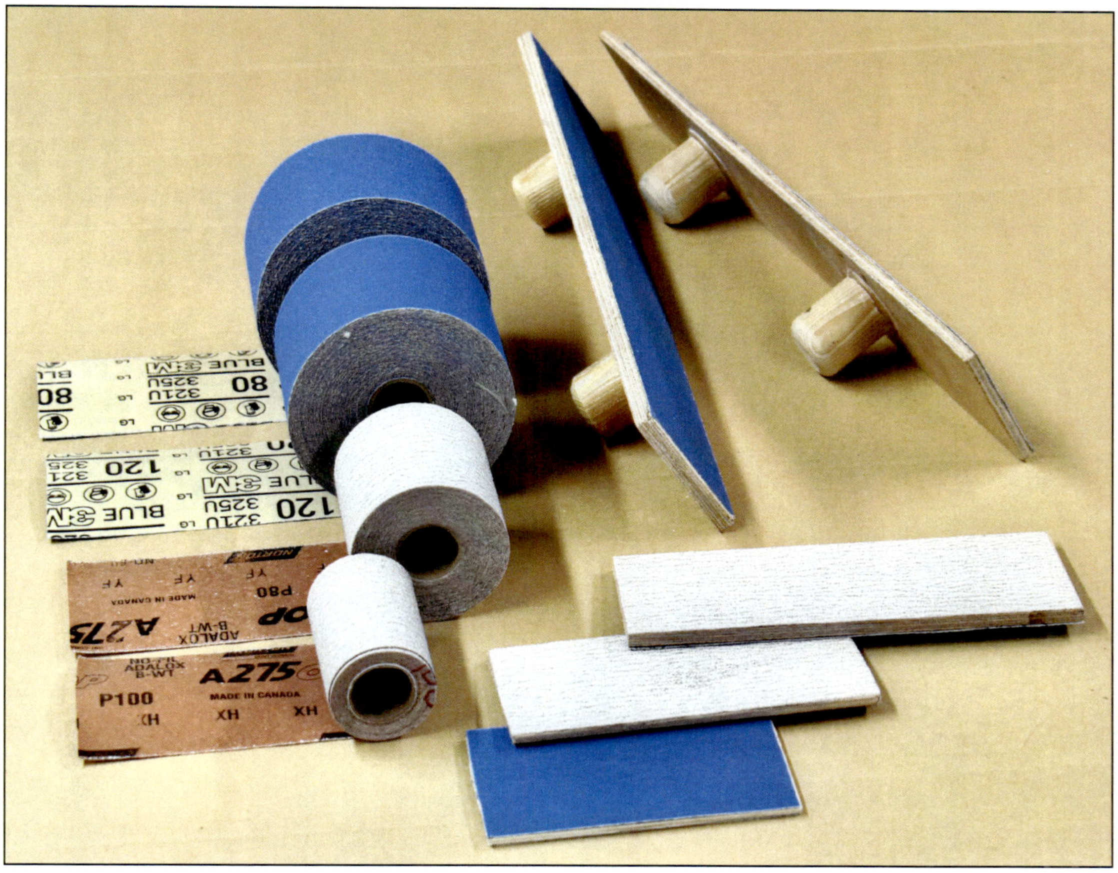

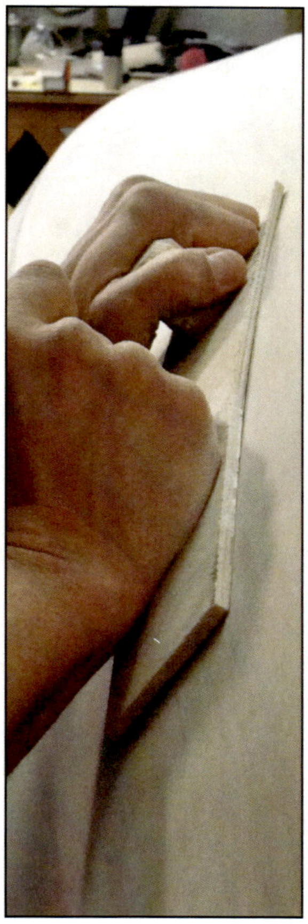

Detail sanding often requires tools that fit the job. Shown are a few of my most-used small tools.
The **tapered sticks** with sandpaper on both sides are for getting into tight places and sanding off drips.
The different size **dowels** are for sanding fillets, etc. They are super handy.
See the line of grit that came off the rolled edge in the smaller photo? That shows where to cut.

If the roll sandpaper is 2 ¾" wide, I make sanding blocks half that wide (1 ⅜") for narrower jobs.
Also, rubber pads can be useful, but they don't produce fair surfaces.

If your sandpaper is clogging, try peeling it off and re-attaching it, knocking it on something, or picking the chunks off with something sharp.
For hand sanding without a block, fold the sandpaper over so it is adhered back-to-back. For economy, one can also spray-glue the back of whole sheets of sandpaper, fold in half, and cut appropriate-sized pieces.

Below is shown the amazing **3M® Scotch-Brite® 07447 abrasive pad**, or what we call the, "3M scrubbie". It is good for removing gloss from surfaces before re-coating or painting.

It can be used for hard to sand areas, such as inside corners and rough surfaces or for prepping to a taped edge.

This stuff's expensive and doesn't stay sharp very long, so cutting pieces of the size needed and replacing when dull is best. For inside corners, pushing a piece of scrubbie on the end of a stick can work well.

ROLLERS & BRUSHES

Like usual, having the right tools makes a big difference. **We use the ⅛" foam rollers** a lot. They are relatively affordable, waste a minimum of epoxy, easily cut to widths needed, and they work!

Cutting the rollers to the minimum needed width saves a lot of epoxy.
They can be cut with a fine-toothed Japanese pull-saw or on a bandsaw.
Different widths require different frames to keep the handle in line with the middle of the roller (3 narrower frames shown below).

For really large glassing jobs or applying thick fill coats, *(see page 26)* a nappy roller can be used. **A short (¼") nap roller can quickly apply a large amount of epoxy**, but will also waste some epoxy, so cutting a 9"wide roller in half and using just half of it seems to work well.

Roller pans are necessary for bigger jobs *(see pallet use page 34)*. A roller pan can hold a large batch of epoxy while it is being applied. Find a smooth (not knobby), flat pan like what is shown. When finished with the job, distribute the residue evenly around the whole inside of the pan, Used pans work best, but the working surface must be kept flat.

We use **foam brushes** from 1" to 4" wide, but mostly use narrower (1" to 2" wide) throw-away brushes.
Throw-away brushes work best with epoxy after being given a haircut *(page 37)*.

Cleaning brushes *(see next page)* can be economical, but they also drop less bristles after cleaning.

Removing goopy roller covers from the frame can be a challenge. Knocking them on the edge of the trash often results in a broken frame.
We like cutting up old bath towels for rags. Using a small piece of towel as shown works well for removing covers. Afterwards, drape the piece of towel over the edge of the trash can. The crustier they get, the better they work.

THINNERS

Thinners are not recommended to be used for anything except cleaning brushes, or cleaning tools that are too sticky to clean with a dry rag or paper towel.
Thinners should NOT be used for surface preparation or for getting epoxy off your skin. *(see "surface preparation", page 40)*

Use acetone if you wish to clean your brushes. Here's how:
Go outside so that you do not trap the harmful fumes in your shop. Pour a small amount of acetone into a cup. Work the brush around in the acetone. **PROTECT YOUR EYES.** Dump the acetone into an old paint tray (this will evaporate quickly), and knock the handle of the brush on the edge of something to remove the acetone. Repeat the process 3 or 4 times. This will use a small amount of acetone and get the brush clean.

Acetone or alcohol can also be used with paper towels to clean the larger (expensive) paper mixing cups.

A good practice is to **not remove the lid from the can of thinner**, but only open it slightly and tip the can to pour a trickle into the paper towel or cup. **Acetone is extremely flammable, so be careful.**

EPOXY SAFETY

Refer to the WEST SYSTEM® literature for more complete safety information. *(Westsytem.com)* Or, safety literature from the epoxy you are using.

Skin contact with uncured epoxy is dangerous and can lead to skin sensitization. When one is sensitized it usually means that working with epoxy is no longer possible, but this is easy to avoid.

Find disposable gloves that you like, and don't get near anything sticky without them on.

Be especially careful not to get *unmixed resin or hardener* on anything but the inside of your mixing cup.

When filling your pumps, have paper towels ready, clean up the drips and deposit directly in the trash.

Multiple trash receptacles around the shop help keep trash where it belongs. (cardboard boxes work well)

Unmixed epoxy in your shop will remain a skin contact hazard and can be spread like germs.

If you get epoxy on your skin, get it off with a waterless skin cleanser (like mechanics use). Warm water and soap is the second choice. NEVER use thinner to clean epoxy off your skin.

Another health hazard is inhaling dust and skin contact from sanding epoxy that has not fully cured. Take extra care with gloves and a respirator if sanding epoxy that hasn't had a few days at room (or warmer) temperatures to cure.

Buy a good respirator. The 3M® half-face choices are light, cost between $20 & $50, and are comfortable, even for long periods.

Use dust filters on your respirator, (such as 3M® #2091 particulate filters) for sanding, sweeping the floor, or mixing fillers into epoxy.

Use charcoal filters for protecting from vapor when coating or glassing larger areas, or painting. A set of charcoal filters cost about $20 (for the 3M® #6001 organic vapor filters). If you remove them from your respirator and **store them in a ziplock bag** when they are not being used, they will last a long time. **Note**: It is important to understand how the three valves on the respirator function and frequently check them to make sure they are seated and working correctly.

A good way to control dust & fumes in the shop is to install an exhaust fan (blowing out) in a wall or window of your shop.

- This fan should be next to your epoxy mixing station (or vice versa) so that you don't need to wear a mask when mixing fillers into epoxy.
- An exhaust fan can provide good cross ventilation by opening a window or door at the other end of the shop.
- If working in a cold climate you may only use the exhaust fan for shop clean-up or sanding large areas, but it would be worth the effort just for that.

Having your shop set up right and being as prepared for each step as possible will help you get the best results when working with epoxy. Enjoying your project is a good measurement of success.

www.ingramcontent.com/pod-product-compliance
Lightning Source LLC
LaVergne TN
LVRC080247220926
848482LV00042B/349

* 9 7 9 8 9 9 6 5 3 7 9 0 7 *